MODELS OF DNA SEQUENCE EVOLUTION

MODELS OF
DNA SEQUENCE EVOLUTION

DR. R.N. SHUKLA

Oxford Book Company
Jaipur ▪ New Delhi

Published By

Oxford Book Company

Regd. Off. "Bony Residency", Gate No. 2
Opp. Tilak Public School, Vishweswariya Nagar,
Gopalpura Road, Jaipur - 302018, India
Ph: 0141 - 2761280, Telefax: 2761381
email - oxfordbookjpr@gmail.com
website - oxfordbookcompany.com

Delhi Office: 102, 1st Floor, Satyam House,
4327/3, Ansari Road, Darya Ganj,
New Delhi - 110002, India
Ph: 011-45652440

Distributed By

KEYA INTERNATIONAL PRESS LLP
E-mail: keyainternationalpressllp@gmail.com

Published in India, 2025

ISBN : 9789391096519

Typeset by Shivangi Computers, Jaipur

Printed and bound in India by Thomson Press Ltd.,
New Delhi

Contents

Preface

The DNA sample is divided into four separate sequencing reactions, containing all four of the standard deoxynucleotides and the DNA polymerase. To each reaction is added only one of the four dideoxynucleotides which are the chain-terminating nucleotides, lacking a 3'-OH group required for the formation of a phosphodiester bond between two nucleotides, thus terminating DNA strand extension and resulting in DNA fragments of varying length. The newly synthesized and labelled DNA fragments are heat denatured, and separated by size by gel electrophoresis on a denaturing polyacrylamide-urea gel with each of the four reactions run in one of four individual lanes; the DNA bands are then visualized byautoradiography or UV light, and the DNA sequence can be directly read off the X-ray film or gel image. In the image on the right, X-ray film was exposed to the gel, and the dark bands correspond to DNA fragments of different lengths. A dark band in a lane indicates a DNA fragment that is the result of chain termination after incorporation of a dideoxynucleotide. The relative positions of the different bands among the four lanes are then used to read the DNA sequence.

DNA sequencing includes several methods and technologies that are used for determining the order of the nucleotide bases—adenine, guanine, cytosine, and thymine—in a molecule of DNA. Knowledge of DNA sequences has become indispensable for basic biological research, other research branches utilizing DNA sequencing, and in numerous applied fields such as diagnostic, biotechnology, forensic biology and biological systematics. The advent of DNA sequencing has significantly accelerated biological research and discovery. The rapid speed of sequencing attained with modern DNA sequencing technology has been instrumental in the sequencing of the human genome, in the Human Genome Project.

Related projects, often by scientific collaboration across continents, have generated the complete DNA sequences of many animal, plant, and microbial genomes. The first DNA sequences were obtained in the early 1970s by academic researchers using laborious methods based on two-dimensional chromatography. Following the development of dye-based sequencing methods with automated analysis, DNA sequencing has become easier and orders of magnitude faster.

The book will convey the excitement of the continuing process of discoveries and the strategies developing in genetic engineering.

— *Dr. R.N. Shukla*

Chapter 1

DNA Nanotechnology

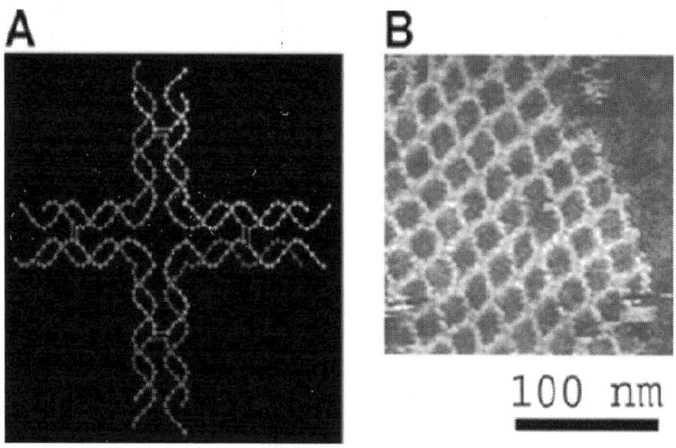

Fig. 1:

The DNA structure at left (schematic shown) will self-assemble into the structure visualized by atomic force microscopy at right. DNA nanotechnology is the field that seeks to design nanoscale structures using the molecular recognitionproperties of DNA molecules. Image from Strong, 2004.

DNA nanotechnology uses the unique molecular recognition properties of DNA and other nucleic acids to create self-assembling branched DNA complexes with useful properties. DNA is thus used as a structural material rather than as a carrier of biological information. This has led to the creation of two-dimensional periodic lattices (both tile-based as well as using the "DNA origami" method) as well as three-dimensional structures in the shapes of polyhedra. Nanomechanical devices and algorithmic self-assembly have also been demonstrated, and these DNA structures have been used to template the arrangement of other molecules such as gold nanoparticles and streptavidin proteins.

History and Anthropology

Because DNA collects mutations over time, which are then inherited, it contains historical information, and, by comparing DNA sequences, geneticists can infer the evolutionary history of organisms, their phylogeny. This field of phylogenetics is a powerful tool in evolutionary biology. If DNA sequences within a species are compared, population geneticists can learn the history of particular populations. This can be used in studies ranging from ecological genetics to anthropology; For example, DNA evidence is being used to try to identify the Ten Lost Tribes of Israel.

DNA has also been used to look at modern family relationships, such as establishing family relationships between the descendants of Sally Hemings and Thomas Jefferson. This usage is closely related to the use of DNA in criminal investigations detailed above. Indeed, some criminal investigations have been solved when DNA from crime scenes has matched relatives of the guilty individual.

History of DNA Research

Fig. 2: James D. Watson and Francis Crick(right), co-originators of the double-helix model, with Maclyn McCarty (left).

DNA was first isolated by the Swiss physician Friedrich Miescher who, in 1869, discovered a microscopic substance in the pus of discarded surgical bandages.

As it resided in the nuclei of cells, he called it "nuclein". In 1878, Albrecht Kossel isolated the non-protein component of "nuclein", nucleic acid, and later isolated its five primary nucleobases.

In 1919, Phoebus Levene identified the base, sugar and phosphate nucleotide unit. Levene suggested that DNA consisted of a string of nucleotide units linked together through the phosphate groups. However, Levene thought the chain was short and the bases repeated in a fixed order. In 1937 William Astbury produced the first X-ray diffraction patterns that showed that DNA had a regular structure.

Fig. 3: Rosalind Franklin used X-ray crystallography to help visualize the structure of DNA.

In 1927 Nikolai Koltsov proposed that inherited traits would be inherited via a "giant hereditary molecule" which would be made up of "two mirror strands that would replicate in a semi-conservative fashion using each strand as a template". In 1928, Frederick Griffith discovered that traits of the "smooth" form of the Pneumococcus could be transferred to the "rough" form of the same bacteria by mixing killed "smooth" bacteria with the live "rough" form.

This system provided the first clear suggestion that DNA carries genetic information—the Avery–MacLeod–McCarty experiment—when Oswald Avery, along with coworkers Colin MacLeod and Maclyn McCarty, identified DNA as the transforming principle in 1943. DNA's role in heredity was confirmed in 1952, when Alfred Hershey and Martha Chase in the Hershey–Chase experiment showed that DNA is the genetic material of the T2 phage.

In 1953, James D. Watson and Francis Crick suggested what is now accepted as the first correct double-helix model of DNA structure in the journal Nature.

Their double-helix, molecular model of DNA was then based on a single X-ray diffraction image (labeled as "Photo 51") taken by Rosalind Franklin and Raymond Gosling in May 1952, as well as the information that the DNA bases are paired — also obtained through private communications from Erwin Chargaff in the previous years.

Chargaff's rules played a very important role in establishing double-helix configurations for B-DNA as well as A-DNA.

Experimental evidence supporting the Watson and Crick model were published in a series of five articles in the same issue of Nature. Of these, Franklin and Gosling's paper was the first publication of their own X-ray diffraction data and original analysis method that partially supported the Watson and Crick model;this issue also contained an article on DNA structure by Maurice Wilkins and two of his colleagues, whose analysis and in vivo B-DNA X-ray patterns also supported the presence in vivo of the double-helical DNA configurations as proposed by Crick and Watson for their double-helix molecular model of DNA in the previous two pages of Nature. In 1962, after Franklin's death, Watson, Crick, and Wilkins jointly received the Nobel Prize in Physiology or Medicine.

However, Nobel rules of the time allowed only living recipients, but a vigorous debate continues on who should receive credit for the discovery.

In an influential presentation in 1957, Crick laid out the central dogma of molecular biology, which foretold the relationship between DNA, RNA, and proteins, and articulated the "adaptor hypothesis".

Final confirmation of the replication mechanism that was implied by the double-helical structure followed in 1958 through the Meselson–Stahl experiment.

Further work by Crick and coworkers showed that the genetic code was based on non-overlapping triplets of bases, called codons, allowing Har Gobind Khorana, Robert W. Holley and Marshall Warren Nirenberg to decipher the genetic code. These findings represent the birth of molecular biology.

RNA

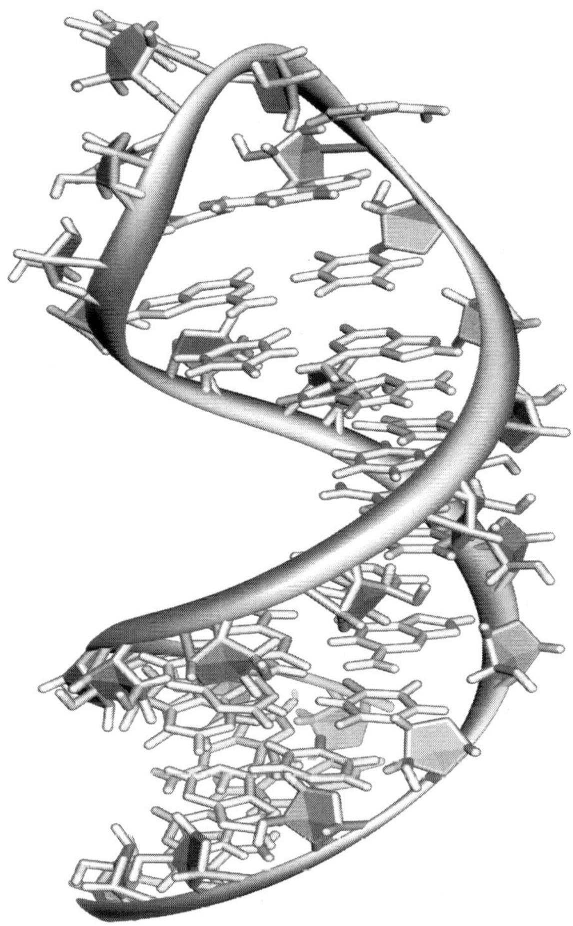

Fig. 4: A hairpin loop from a pre-mRNA. Highlighted are the nucleobases (green) and the ribose-phosphate backbone (blue).

Ribonucleic acid (English pronunciation: /rajbu.njuÐÌklej.hk Èæsjd), or RNA, is one of the three major macromolecules (along with DNA and proteins) that are essential for all known forms of life.

Like DNA, RNA is made up of a long chain of components called nucleotides.

Each nucleotide consists of a nucleobase (sometimes called a nitrogenous base), a ribose sugar, and a phosphate group. The sequence of nucleotides allows RNA to encode genetic information. All cellular organisms use messenger RNA(mRNA) to carry the genetic information that directs the synthesis of proteins.

In addition, some viruses use RNA instead of DNA as their genetic material; perhaps a reflection of the suggested key role of RNA in the evolutionary history of life on Earth.

Like proteins, some RNA molecules play an active role in cells by catalyzing biological reactions, controlling gene expression, or sensing and communicating responses to cellular signals.

One of these active processes is protein synthesis, a universal function whereby mRNA molecules direct the assembly of proteins on ribosomes.

This process uses transfer RNA (tRNA) molecules to deliver amino acids to the ribosome, where ribosomal RNA (rRNA) links amino acids together to form proteins. In 2011, it was proved that the methylation of mRNA has a critical role in human energy homeostasis. This opens up the field of RNA epigenetics.

The chemical structure of RNA is very similar to that of DNA, with two differences: (a) RNA contains the sugar ribose, while DNA contains the slightly different sugar deoxyribose (a type of ribose that lacks one oxygen atom), and (b) RNA has the nucleobase uracil while DNA contains thymine. Uracil and thymine have similar base-pairing properties.

Unlike DNA, most RNA molecules are single-stranded. Single-stranded RNA molecules adopt very complex three-dimensional structures, since they are not restricted to the repetitive double-helical form of double-stranded DNA.

RNA is made within living cells by RNA polymerases, enzymes that act to copy a DNA or RNA template into a new RNA strand through processes known as transcription or RNA replication, respectively.

Comparison with DNA

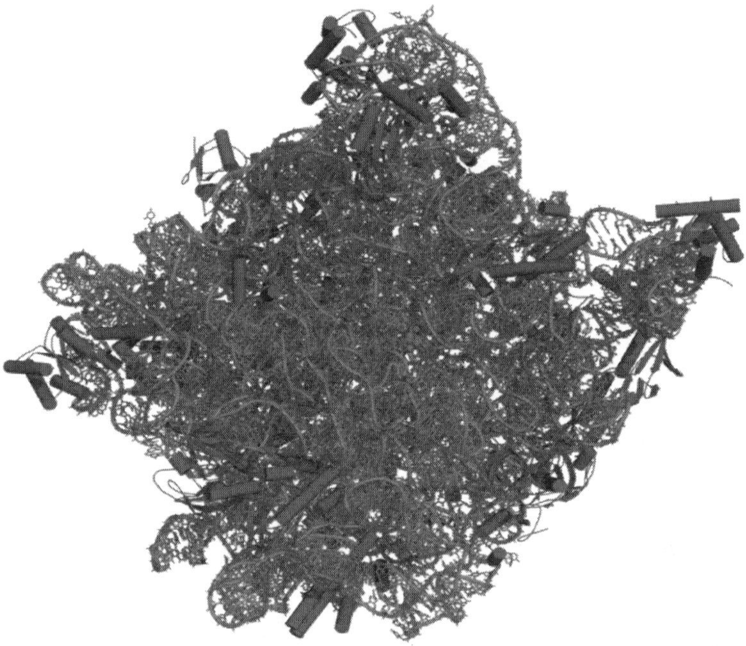

Fig. 5: Three-dimensional representation of the50S ribosomal subunit. RNA is in ochre, protein in blue. The active site is in the middle (red).

RNA and DNA are both nucleic acids, but differ in three main ways. First, unlike double-stranded DNA, RNA is a single-stranded molecule in many of its biological roles and has a much shorter chain of nucleotides. Second, while DNA contains deoxyribose, RNA contains ribose (in deoxyribose there is no hydroxyl group attached to the pentose ring in the 2' position).

These hydroxyl groups make RNA less stable than DNA because it is more pronc tohydrolysis. Third, the complementary base to adenine is not thymine, as it is in DNA, but rather uracil, which is an unmethylated form of thymine.

Like DNA, most biologically active RNAs, including mRNA, tRNA, rRNA, snRNAs, and other non-coding RNAs, contain self-complementary sequences that allow parts of the RNA to fold and pair with itself to form double helices. Analysis of these RNAs has revealed that they are highly structured. Unlike DNA, their structures do not consist of long double helices but rather collections of short helices packed together into structures akin to proteins.

In this fashion, RNAs can achieve chemical catalysis, like enzymes. For instance, determination of the structure of the ribosome—an enzyme that catalyzes peptide bond formation—revealed that its active site is composed entirely of RNA.

Structure

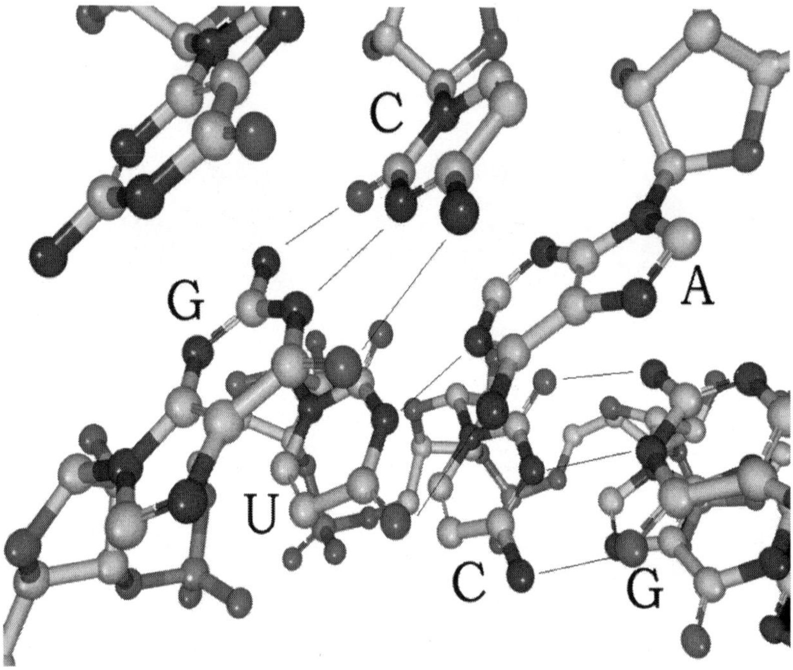

Fig. 6: Watson-Crick base pairs in a siRNA(hydrogen atoms are not shown)

Each nucleotide in RNA contains a ribose sugar, with carbons numbered 1' through 5'. A base is attached to the 1' position, in general, adenine (A), cytosine(C), guanine (G), or uracil (U). Adenine and guanine are purines, cytosine, and uracil are pyrimidines. A phosphate group is attached to the 3' position of one ribose and the 5' position of the next.

The phosphate groups have a negative charge each at physiological pH, making RNA a charged molecule (polyanion). The bases may form hydrogen bonds between cytosine and guanine, between adenine and uracil and between guanine and uracil.

However, other interactions are possible, such as a group of adenine bases binding to each other in a bulge, or the GNRA tetraloop that has a guanine–adenine base-pair.

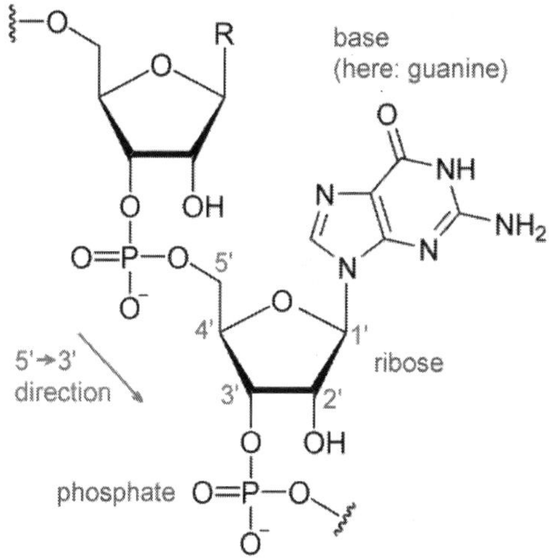

Fig. 7:

Chemical Structure of RNA

An important structural feature of RNA that distinguishes it from DNA is the presence of a hydroxyl group at the 2' position of the ribose sugar. The presence of this functional group causes the helix to adopt the A-form geometry rather than the B-form most commonly observed in DNA.

This results in a very deep and narrow major groove and a shallow and wide minor groove. A second consequence of the presence of the 2'-hydroxyl group is that in conformationally flexible regions of an RNA molecule (that is, not involved in formation of a double helix), it can chemically attack the adjacent phosphodiester bond to cleave the backbone.

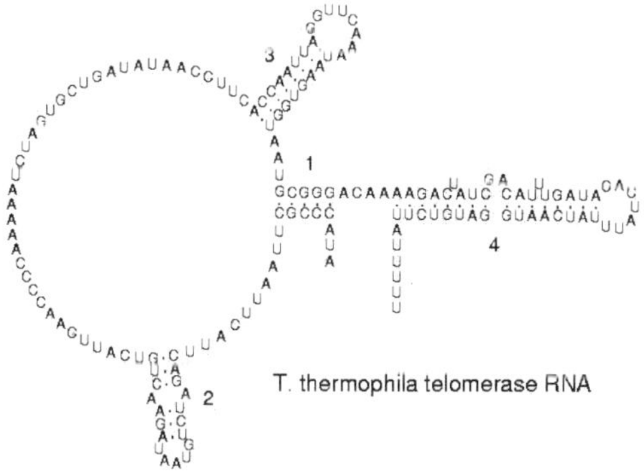

Fig. 8: Secondary structure of a telomerase RNA.

RNA is transcribed with only four bases (adenine, cytosine, guanine and uracil), but these bases and attached sugars can be modified in numerous ways as the RNAs mature.

Pseudouridine (Ø), in which the linkage between uracil and ribose is changed from a C–N bond to a C–C bond, and ribothymidine (T) are found in various places (the most notable ones being in the TØC loop of tRNA).

Another notable modified base is hypoxanthine, a deaminated adenine base whose nucleoside is called inosine (I). Inosine plays a key role in the wobble hypothesisof the genetic code.

There are nearly 100 other naturally occurring modified nucleosides, of which pseudouridine and nucleosides with 2'-O-methylribose are the most common.

The specific roles of many of these modifications in RNA are not fully understood. However, it is notable that, in ribosomal RNA, many of the post-transcriptional modifications occur in highly functional regions, such as the peptidyl transferase center and the subunit interface, implying that they are important for normal function.

The functional form of single stranded RNA molecules, just like proteins, frequently requires a specific tertiary structure. The scaffold for this structure is provided by secondary structural elements that are hydrogen bonds within the molecule.

This leads to several recognizable "domains" of secondary structure like hairpin loops, bulges, and internal loops. Since RNA is charged, metal ions such as Mg are needed to stabilise many secondary and tertiary structures.

Synthesis

Synthesis of RNA is usually catalyzed by an enzyme—RNA polymerase—using DNA as a template, a process known as transcription. Initiation of transcription begins with the binding of the enzyme to a promoter sequence in the DNA (usually found "upstream" of a gene).

The DNA double helix is unwound by the helicase activity of the enzyme. The enzyme then progresses along the template strand in the 3' to 5' direction, synthesizing a complementary RNA molecule with elongation occurring in the 5' to 3' direction.

The DNA sequence also dictates where termination of RNA synthesis will occur.

RNAs are often modified by enzymes after transcription. For example, a poly(A) tail and a 5' cap are added to eukaryotic pre-mRNA and introns are removed by the spliceosome.

There are also a number of RNA-dependent RNA polymerases that use RNA as their template for synthesis of a new strand of RNA. For instance, a number of RNA viruses (such as poliovirus) use this type of enzyme to replicate their genetic material.

Also, RNA-dependent RNA polymerase is part of the RNA interference pathway in many organisms.

Types of RNA

Overview

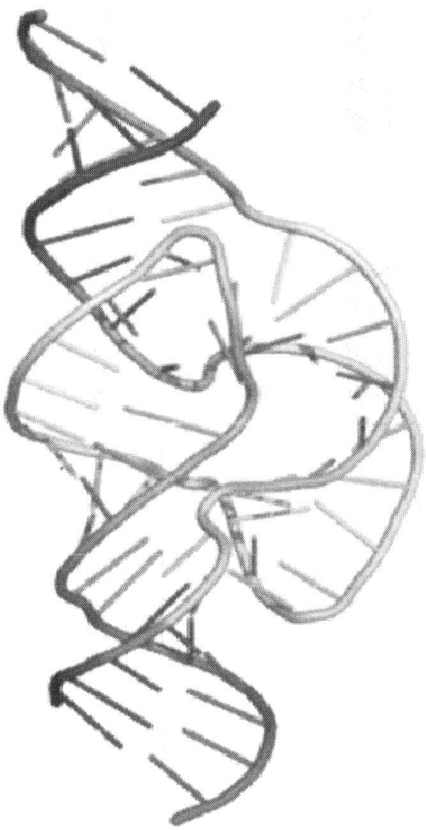

Fig. 9: Structure of a hammerhead ribozyme, a ribozyme that cuts RNA

Messenger RNA (mRNA) is the RNA that carries information from DNA to the ribosome, the sites of protein synthesis (translation) in the cell.

The coding sequence of the mRNA determines the amino acid sequence in the protein that is produced. Many RNAs do not code for protein however (about 97% of the transcriptional output is non-protein-coding in eukaryotes).

These so-called non-coding RNAs ("ncRNA") can be encoded by their own genes (RNA genes), but can also derive from mRNA introns. The most prominent examples of non-coding RNAs are transfer RNA (tRNA) and ribosomal RNA (rRNA), both of which are involved in the process of translation.

There are also non-coding RNAs involved in gene regulation, RNA processing and other roles. Certain RNAs are able to catalyse chemical reactions such as cutting and ligating other RNA molecules, and the catalysis of peptide bond formation in the ribosome; these are known as ribozymes.

In Translation

Messenger RNA (mRNA) carries information about a protein sequence to the ribosomes, the protein synthesis factories in the cell. It is coded so that every three nucleotides (a codon) correspond to one amino acid.

In eukaryotic cells, once precursor mRNA (pre-mRNA) has been transcribed from DNA, it is processed to mature mRNA. This removes its introns—non-coding sections of the pre-mRNA. The mRNA is then exported from the nucleus to the cytoplasm, where it is bound to ribosomes and translated into its corresponding protein form with the help of tRNA.

In prokaryotic cells, which do not have nucleus and cytoplasm compartments, mRNA can bind to ribosomes while it is being transcribed from DNA. After a certain amount of time the message degrades into its component nucleotides with the assistance of ribonucleases.

Transfer RNA (tRNA) is a small RNA chain of about 80 nucleotides that transfers a specific amino acid to a growing polypeptide chain at the ribosomal site of protein synthesis during translation.

It has sites for amino acid attachment and an anticodon region for codon recognition that binds to a specific sequence on the messenger RNA chain through hydrogen bonding.

Ribosomal RNA (rRNA) is the catalytic component of the ribosomes. Eukaryotic ribosomes contain four different rRNA molecules: 18S, 5.8S, 28S and 5S rRNA.

Three of the rRNA molecules are synthesized in the nucleolus, and one is synthesized elsewhere. In the cytoplasm, ribosomal RNA and protein combine to form a nucleoprotein called a ribosome. The ribosome binds mRNA and carries out protein synthesis. Several ribosomes may be attached to a single mRNA at any time. Nearly all the RNA found in a typical eukaryotic cell is rRNA.

Transfer-messenger RNA (tmRNA) is found in many bacteria and plastids. It tags proteins encoded by mRNAs that lack stop codons for degradation and prevents the ribosome from stalling.

Regulatory RNAs

Several types of RNA can downregulate gene expression by being complementary to a part of an mRNA or a gene's DNA. MicroRNAs (miRNA; 21-22 nt) are found in eukaryotes and act through RNA interference (RNAi), where an effector complex of miRNA and enzymes can cleave complementary mRNA, block the mRNA from being translated, or accelerate its degradation.

While small interfering RNAs (siRNA; 20-25 nt) are often produced by breakdown of viral RNA, there are also endogenous sources of siRNAs.

siRNAs act through RNA interference in a fashion similar to miRNAs. Some miRNAs and siRNAs can cause genes they target to be methylated, thereby decreasing or increasing transcription of those genes.

Animals have Piwi-interacting RNAs (piRNA; 29-30 nt) which are active in germline cells and are thought to be a defense against transposons and play a role in gametogenesis.

Many prokaryotes have CRISPR RNAs, a regulatory system similar to RNA interference. Antisense RNAs are widespread; most downregulate a gene, but a few are activators of transcription.One way antisense RNA can act is by binding to an mRNA, forming double-stranded RNA that is enzymatically degraded.

There are many long noncoding RNAs that regulate genes in eukaryotes, one such RNA is Xist, which coats one X chromosome in female mammals and inactivates it.

An mRNA may contain regulatory elements itself, such as riboswitches, in the 5' untranslated region or 3' untranslated region; these cis-regulatory elements regulate the activity of that mRNA. The untranslated regions can also contain elements that regulate other genes.

In RNA Processing

Many RNAs are involved in modifying other RNAs. Introns are spliced out of pre-mRNA by spliceosomes, which contain several small nuclear RNAs(snRNA), or the introns can be ribozymes that are spliced by themselves. RNA can also be altered by having its nucleotides modified to other nucleotides than A, C, G and U. In eukaryotes, modifications of RNA nucleotides are generally directed by small nucleolar

RNAs (snoRNA; 60-300 nt),found in the nucleolus and cajal bodies. snoRNAs associate with enzymes and guide them to a spot on an RNA by basepairing to that RNA. These enzymes then perform the nucleotide modification. rRNAs and tRNAs are extensively modified, but snRNAs and mRNAs can also be the target of base modification. RNA can also be methylated.

RNA Genomes

Like DNA, RNA can carry genetic information. RNA viruses have genomes composed of RNA which encodes a number of proteins. The viral genome is replicated by some of those proteins, while other proteins protect the genome as the virus particle moves to a new host cell. Viroids are another group of pathogens, but they consist only of RNA, do not encode any protein and are replicated by a host plant cell's polymerase.

In Reverse Transcription

Reverse transcribing viruses replicate their genomes by reverse transcribing DNA copies from their RNA; these DNA copies are then transcribed to new RNA. Retrotransposons also spread by copying DNA and RNA from one another, and telomerase contains an RNA that is used as template for building the ends of eukaryotic chromosomes.

Double-stranded RNA

Double-stranded RNA (dsRNA) is RNA with two complementary strands, similar to the DNA found in all cells. dsRNA forms the genetic material of some viruses (double-stranded RNA viruses). Double-stranded RNA such as viral RNA or siRNA can trigger RNA interference in eukaryotes, as well as interferon response in vertebrates.

Key Discoveries in RNA Biology

Research on RNA has led to many important biological discoveries and numerous Nobel Prizes. Nucleic acids were discovered in 1868 by Friedrich Miescher, who called the material 'nuclein' since it was found in the nucleus.

It was later discovered that prokaryotic cells, which do not have a nucleus, also contain nucleic acids.

The role of RNA in protein synthesis was suspected already in 1939. Severo Ochoa won the 1959 Nobel Prize in Medicine (shared with Arthur Kornberg) after he discovered an enzyme that can synthesize RNA in the laboratory.

Ironically, the enzyme discovered by Ochoa (polynucleotide phosphorylase) was later shown to be responsible for RNA degradation, not RNA synthesis.

The sequence of the 77 nucleotides of a yeast tRNA was found by Robert W. Holley in 1965, winning Holley the 1968 Nobel Prize in Medicine (shared with Har Gobind Khorana and Marshall Nirenberg).

In 1967, Carl Woese hypothesized that RNA might be catalytic and suggested that the earliest forms of life (self-replicating molecules) could have relied on RNA both to carry genetic information and to catalyze biochemical reactions—an RNA world.

During the early 1970s, retroviruses and reverse transcriptase were discovered, showing for the first time that enzymes could copy RNA into DNA (the opposite of the usual route for transmission of genetic information).

For this work, David Baltimore, Renato Dulbecco and Howard Temin were awarded a Nobel Prize in 1975. In 1976, Walter Fiers and his team determined the first complete nucleotide sequence of an RNA virus genome, that of bacteriophage MS2.

In 1977, introns and RNA splicing were discovered in both mammalian viruses and in cellular genes, resulting in a 1993 Nobel to Philip Sharp and Richard Roberts.

Catalytic RNA molecules (ribozymes) were discovered in the early 1980s, leading to a 1989 Nobel award to Thomas Cech and Sidney Altman.

In 1990 it was found in petunia that introduced genes can silence similar genes of the plant's own, now known to be a result of RNA interference.

At about the same time, 22 nt long RNAs, now called microRNAs,were found to have a role in the development of C. elegans.

Studies on RNA interference gleaned a Nobel Prize for Andrew Fireand Craig Mello in 2006, and another Nobel was awarded for studies on transcription of RNA to Roger Kornberg in the same year.

The discovery of gene regulatory RNAs has led to attempts to develop drugs made of RNA, such as siRNA, to silence genes.

Chapter 2

DNA Sequencing

DNA sequencing includes several methods and technologies that are used for determining the order of the nucleotide bases—adenine, guanine, cytosine, and thymine—in a molecule of DNA.

Knowledge of DNA sequences has become indispensable for basic biological research, other research branches utilizing DNA sequencing, and in numerous applied fields such as diagnostic, biotechnology, forensic biology and biological systematics.

The advent of DNA sequencing has significantly accelerated biological research and discovery. The rapid speed of sequencing attained with modern DNA sequencing technology has been instrumental in the sequencing of the human genome, in the Human Genome Project.

Related projects, often by scientific collaboration across continents, have generated the complete DNA sequences of many animal, plant, and microbial genomes.

The first DNA sequences were obtained in the early 1970s by academic researchers using laborious methods based on two-dimensional chromatography.

Following the development of dye-based sequencing methods with automated analysis, DNA sequencing has become easier and orders of magnitude faster.

History

RNA sequencing was one of the earliest forms of nucleotide sequencing. The major landmark of RNA sequencing is the sequence of the first complete gene and the complete genome of Bacteriophage MS2, identified and published by Walter Fiers and his coworkers at the University of Ghent (Ghent, Belgium), between 1972 and 1976.

Prior to the development of rapid DNA sequencing methods in the early 1970s by Frederick Sanger at the University of Cambridge, in England and Walter Gilbert and Allan Maxam at Harvard, a number of laborious methods were used. For instance, in 1973, Gilbert and Maxam reported the sequence of 24 basepairs using a method known as wandering-spot analysis.

The chain-termination method developed by Sanger and coworkers in 1977 soon became the method of choice, owing to its relative ease and reliability. It involves separating DNA bases from different DNA fragments.

Maxam–Gilbert Sequencing

In 1976–1977, Allan Maxam and Walter Gilbert developed a DNA sequencing method based on chemical modification of DNA and subsequent cleavage at specific bases. Although Maxam and Gilbert published their chemical sequencing method two years after the ground-breaking paper of Sanger and Coulson on plus-minus sequencing, Maxam–Gilbert sequencing rapidly became more popular, since purified DNA could be used directly, while the initial Sanger method required that each read start be cloned for production of single-stranded DNA.

However, with the improvement of the chain-termination method (see below), Maxam-Gilbert sequencing has fallen out of favour due to its technical complexity prohibiting its use in standard molecular biology kits, extensive use of hazardous chemicals, and difficulties with scale-up.

The method requires radioactive labeling at one 5' end of the DNA (typically by a kinase reaction using gamma-P ATP) and purification of the DNA fragment to be sequenced. Chemical treatment generates breaks at a small proportion of one or two of the four nucleotide bases in each of four reactions (G, A+G, C, C+T).

For example, the purines (A+G) are depurinated using formic acid, the guanines (and to some extent the adenines) are methylated by dimethyl sulfate, and the pyrimidines (C+T) are methylated using hydrazine.

The addition of salt (sodium chloride) to the hydrazine reaction inhibits the methylation of thymine for the C-only reaction. The modified DNAs are then cleaved by hot piperidine at the position of the modified base. The concentration of the modifying chemicals is controlled to introduce on average one modification per DNA molecule.

Thus a series of labeled fragments is generated, from the radiolabeled end to the first "cut" site in each molecule.

The fragments in the four reactions are electrophoresed side by side in denaturing acrylamide gels for size separation. To visualize the fragments, the gel is exposed to X-ray film for autoradiography, yielding a series of dark bands each corresponding to a radiolabeled DNA fragment, from which the sequence may be inferred.

Also sometimes known as "chemical sequencing", this method led to the Methylation Interference Assay used to map DNA-binding sites for DNA-binding proteins.

Chain-termination Methods

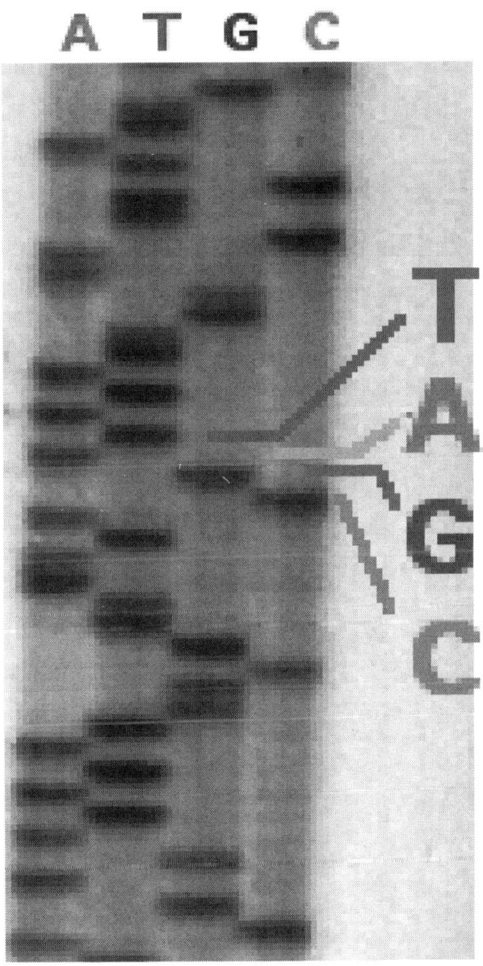

Fiog. 1: Part of a radioactively labelled sequencing gel

Because the chain-terminator method (or Sanger method after its developer Frederick Sanger) is more efficient and uses fewer toxic chemicals and lower amounts of radioactivity than the method of Maxam and Gilbert, it rapidly became the method of choice. The key principle of the Sanger method was the use of dideoxy nucleotide triphosphates (ddNTPs) as DNA chain terminators.

The classical chain-termination method requires a single-stranded DNA template, a DNA primer, a DNA polymerase, normal deoxynucleotidetriphosphates (dNTPs), and modified nucleotides (dideoxyNTPs) that terminate DNA strand elongation. These ddNTPs will also be radioactively or fluorescently labelled for detection in automated sequencing machines.

The DNA sample is divided into four separate sequencing reactions, containing all four of the standard deoxynucleotides (dATP, dGTP, dCTP and dTTP) and the DNA polymerase. To each reaction is added only one of the four dideoxynucleotides (ddATP, ddGTP, ddCTP, or ddTTP) which are the chain-terminating nucleotides, lacking a 3'-OH group required for the formation of a phosphodiester bond between two nucleotides, thus terminating DNA strand extension and resulting in DNA fragments of varying length.

The newly synthesized and labelled DNA fragments are heat denatured, and separated by size (with a resolution of just one nucleotide) by gel electrophoresis on a denaturing polyacrylamide-urea gel with each of the four reactions run in one of four individual lanes (lanes A, T, G, C); the DNA bands are then visualized byautoradiography or UV light, and the DNA sequence can be directly read off the X-ray film or gel image.

In the image on the right, X-ray film was exposed to the gel, and the dark bands correspond to DNA fragments of different lengths. A dark band in a lane indicates a DNA fragment that is the result of chain termination after incorporation of a dideoxynucleotide (ddATP, ddGTP, ddCTP, or ddTTP). The relative positions of the different bands among the four lanes are then used to read (from bottom to top) the DNA sequence.

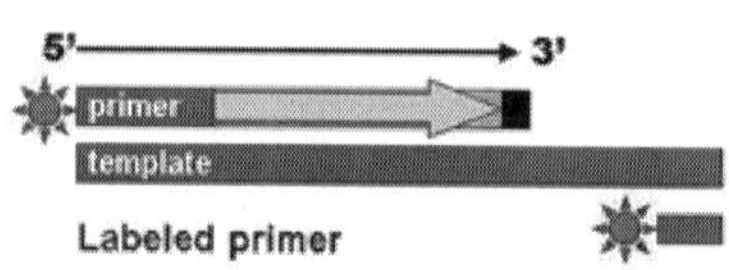

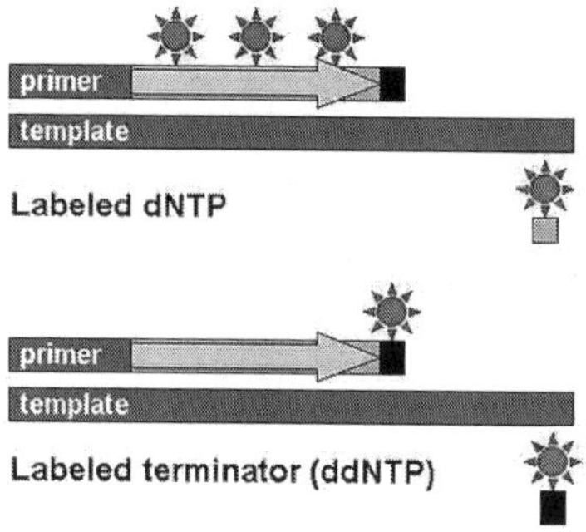

Figure 3: DNA fragments are labelled with a radioactive or fluorescent tag on the primer (1), in the new DNA strand with a labeled dNTP, or with a labeled ddNTP.

Technical variations of chain-termination sequencing include tagging with nucleotides containing radioactive phosphorus forradiolabelling, or using a primer labeled at the 5' end with a fluorescent dye. Dye-primer sequencing facilitates reading in an optical system for faster and more economical analysis and automation. The later development by Leroy Hood and coworkers of fluorescently labeled ddNTPs and primers set the stage for automated, high-throughput DNA sequencing.

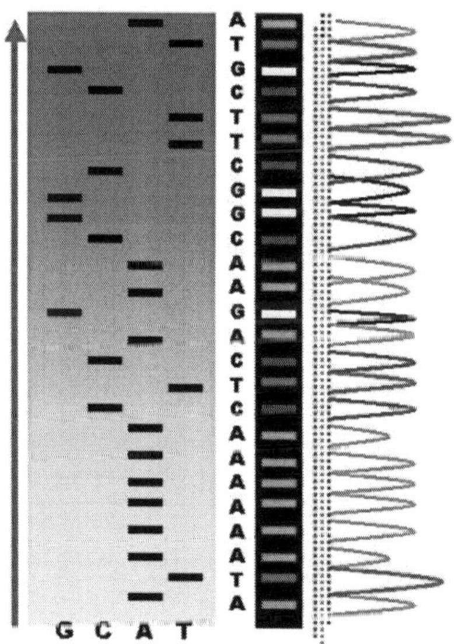

Figure 4: Sequence ladder by radioactive sequencing compared to fluorescent peaks.

Chain-termination methods have greatly simplified DNA sequencing. For example, chain-termination-based kits are commercially available that contain the reagents needed for sequencing, pre-aliquoted and ready to use. Limitations include non-specific binding of the primer to the DNA, affecting accurate read-out of the DNA sequence, and DNA secondary structures affecting the fidelity of the sequence.

Dye-terminator Sequencing

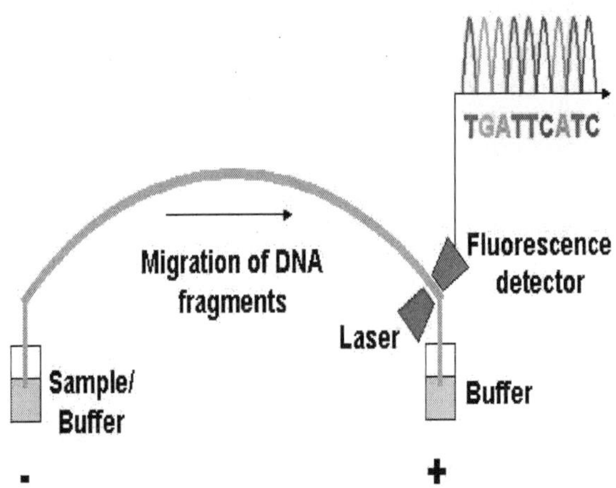

Fig. 5: Capillary electrophoresis.

Dye-terminator sequencing utilizes labelling of the chain terminator ddNTPs, which permits sequencing in a single reaction, rather than four reactions as in the labelled-primer method. In dye-terminator sequencing, each of the four dideoxynucleotide chain terminators is labelled with fluorescent dyes, each of which emit light at differentwavelengths.

Owing to its greater expediency and speed, dye-terminator sequencing is now the mainstay in automated sequencing. Its limitations include dye effects due to differences in the incorporation of the dye-labelled chain terminators into the DNA fragment, resulting in unequal peak heights and shapes in the electronic DNA sequence trace chromatogram after capillary electrophoresis.

This problem has been addressed with the use of modified DNA polymerase enzyme systems and dyes that minimize incorporation variability, as well as methods for eliminating "dye blobs". The dye-terminator sequencing method, along with automated high-throughput DNA sequence analyzers, is now being used for the vast majority of sequencing projects.

Challenges

Common challenges of DNA sequencing include poor quality in the first 15–40 bases of the sequence and deteriorating quality of sequencing traces after 700–900 bases. Base calling software typically gives an estimate of quality to aid in quality trimming.

In cases where DNA fragments are cloned before sequencing, the resulting sequence may contain parts of the cloning vector. In contrast, PCR-based cloning and emerging sequencing technologies based on pyrosequencing often avoid using cloning vectors. Recently, one-step Sanger sequencing (combined amplification and sequencing) methods such as Ampliseq and SeqSharp have been developed that allow rapid sequencing of target genes without cloning or prior amplification.

Current methods can directly sequence only relatively short (300–1000 nucleotides long) DNA fragments in a single reaction. The main obstacle to sequencing DNA fragments above this size limit is insufficient power of separation for resolving large DNA fragments that differ in length by only one nucleotide. In all cases the use of a primer with a free 3' end is essential.

Automation and Sample Preparation

Automated DNA-sequencing instruments (DNA sequencers) can sequence up to 384 DNA samples in a single batch (run) in up to 24 runs a day. DNA sequencers carry out capillary electrophoresis for size separation, detection and recording of dye fluorescence, and data output as fluorescent peak tracechromatograms.

Sequencing reactions by thermocycling, cleanup and re-suspension in a buffer solution before loading onto the sequencer are performed separately.

A number of commercial and non-commercial software packages can trim low-quality DNA traces automatically. These programs score the quality of each peak and remove low-quality base peaks (generally located at the ends of the sequence). The accuracy of such algorithms is below visual examination by a human operator, but sufficient for automated processing of large sequence data sets.

Amplification and Clonal Selection

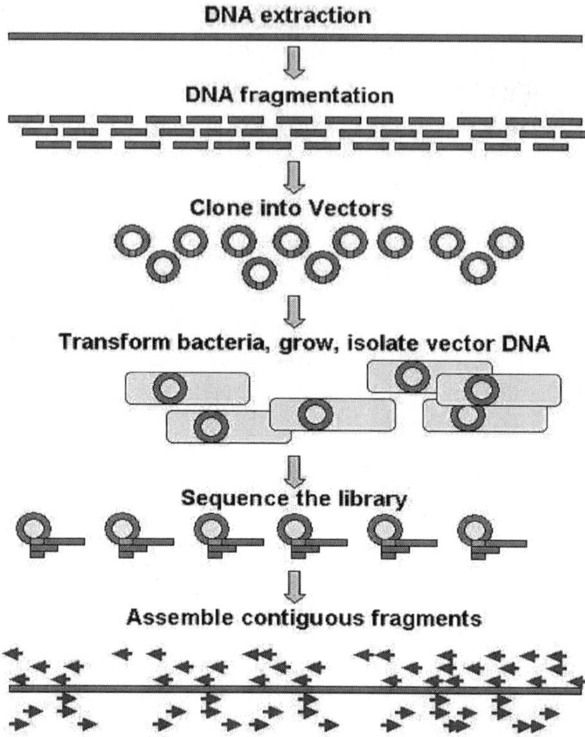

Fig. 6: Genomic DNA is fragmented into random pieces and cloned as a bacterial library. DNA from individual bacterial clones is sequenced and the sequence is assembled by using overlapping DNA regions.

Large-scale sequencing often aims at sequencing very long DNA pieces, such as whole chromosomes, although large scale sequencing can also be used to generate very large numbers of short sequences, such as found in phage display. For longer targets, such as chromosomes, common approaches consist of cutting (with restriction enzymes) or shearing (with mechanical forces) large DNA fragments into shorter DNA fragments.

The fragmented DNA is cloned into aDNA vector, and amplified in Escherichia coli. Short DNA fragments purified from individual bacterial colonies are individually sequenced and assembled electronically into one long, contiguous sequence.

This method does not require any pre-existing information about the sequence of the DNA and is referred to as de novo sequencing. Gaps in the assembled sequence may be filled by primer walking. The different strategies have different tradeoffs in speed and accuracy; shotgun methods are often used for sequencing large genomes, but its assembly is complex and difficult, particularly with sequence repeats often causing gaps in genome assembly.

Most sequencing approaches use an in vitro cloning step to amplify individual DNA molecules, because their molecular detection methods are not sensitive enough for single molecule sequencing. Emulsion PCR isolates individual DNA molecules along with primer-coated beads in aqueous droplets within an oil phase.

Polymerase chain reaction (PCR) then coats each bead with clonal copies of the DNA molecule followed by immobilization for later sequencing. Emulsion PCR is used in the methods by Marguilis et al. (commercialized by 454 Life Sciences), Shendure and Porreca et al. (also known as "Polony sequencing") and SOLiD sequencing, (developed by Agencourt, now Applied Biosystems).

Another method for in vitro clonal amplification is bridge PCR, where fragments are amplified upon primers attached to a solid surface, used in the IlluminaGenome Analyzer. Single-molecule methods, such as that developed by Stephen Quake's laboratory (later commercialized by Helicos) is an exception:

> *it uses bright fluorophores and laser excitation to detect base addition events from individual DNA molecules fixed to a surface, eliminating the need for molecular amplification.*

High-throughput Sequencing

The high demand for low-cost sequencing has driven the development of high-throughput sequencing technologies that parallelize the sequencing process, producing thousands or millions of sequences at once. High-throughput sequencing technologies are intended to lower the cost of DNA sequencing beyond what is possible with standard dye-terminator methods.

The first of the "next-generation" sequencing technologies, MPSS was developed in the 1990s at Lynx Therapeutics, a company founded in 1992 by Sydney Brenner and Sam Eletr. MPSS was a bead-based method that used a complex approach of adapter ligation followed by adapter decoding, reading the sequence in increments of four nucleotides; this method made it susceptible to sequence-specific bias or loss of specific sequences.

Because the technology was so complex, MPSS was only performed 'in-house' by Lynx Therapeutics and no machines were sold; when the merger with Solexa later led to the development of sequencing-by-synthesis, a more simple approach with numerous advantages, MPSS became obsolete.

However, the essential properties of the MPSS output were typical of later "next-gen" data types, including hundreds of thousands of short DNA sequences. In the case of MPSS, these were typically used for sequencing cDNA for measurements of gene expression levels. Lynx Therapeutics merged with Solexa in 2004, and this company was later purchased by Illumina.

Polony Sequencing

Polony sequencing, developed in the laboratory of George Church at Harvard, was among the first next-generation sequencing systems used to sequence a full genome in 2005.

It combined an in vitro paired-tag library with emulsion PCR, an automated microscope, and ligation-based sequencing chemistry to sequence an E. coli genome at an accuracy of > 99.9999% and a cost approximately 1/10 that of Sanger sequencing. The technology was licensed to Agencourt Biosciences, subsequently spun out into Agencourt Personal Genomics, and ultimately incorporated into the Applied Biosystems SOLiD platform.

Pyrosequencing

A parallelized version of pyrosequencing was developed by 454 Life Sciences, which has since been acquired by Roche Diagnostics. The method amplifies DNA inside water droplets in an oil solution (emulsion PCR), with each droplet containing a single DNA template attached to a single primer-coated bead that then forms a clonal colony.

The sequencing machine contains many picolitre-volume wells each containing a single bead and sequencing enzymes. Pyrosequencing uses luciferase to generate light for detection of the individual nucleotides added to the nascent DNA, and the combined data are used to generate sequence read-outs. This technology provides intermediate read length and price per base compared to Sanger sequencing on one end and Solexa and SOLiD on the other.

Illumina (Solexa) Sequencing

Solexa, now part of Illumina, developed a sequencing technology based on reversible dye-terminators. DNA molecules are first attached to primers on a slide and amplified so that local clonal colonies are formed (bridge amplification).

Four types of ddNTPs are added, and non-incorporated nucleotides are washed away. Unlike pyrosequencing, the DNA can only be extended one nucleotide at a time. A camera takes images of the fluorescently labeled nucleotides, then the dye along with the terminal 3' blocker is chemically removed from the DNA, allowing the next cycle.

Solid Sequencing

Applied Biosystems' SOLiD technology employs sequencing by ligation. Here, a pool of all possible oligonucleotides of a fixed length are labeled according to the sequenced position. Oligonucleotides are annealed and ligated; the preferential ligation by DNA ligase for matching sequences results in a signal informative of the nucleotide at that position. Before sequencing, the DNA is amplified by emulsion PCR. The resulting bead, each containing only copies of the same DNA molecule, are deposited on a glass slide. The result is sequences of quantities and lengths comparable to Illumina sequencing.

Ion Semiconductor Sequencing

Ion Torrent Systems Inc. developed a system based on using standard sequencing chemistry, but with a novel, semiconductor based detection system. This method of sequencing is based on the detection of hydrogen ions that are released during the polymerisation of DNA, as opposed to the optical methods used in other sequencing systems. A microwell containing a template DNA strand to be sequenced is flooded with a single type of nucleotide. If the introduced nucleotide is complementary to the leading template nucleotide it is incorporated into the growing complementary strand.

This causes the release of a hydrogen ion that triggers a hypersensitive ion sensor, which indicates that a reaction has occurred. If homopolymer repeats are present in the template sequence multiple nucleotides will be incorporated in a single cycle. This leads to a corresponding number of released hydrogens and a proportionally higher electronic signal.

DNA Nanoball Sequencing

DNA nanoball sequencing is a type of high throughput sequencing technology used to determine the entire genomic sequence of an organism. The company Complete Genomics uses this technology to sequence samples that researchers submit from several projects.

The method uses rolling circle replication to amplify small fragments of genomic DNA into DNA nanoballs. Unchained sequencing by ligation is then used to determine the nucleotide sequence. This method of DNA sequencing allows large numbers of DNA nanoballs to be sequenced per run and at low reagent costs compared to other next generation sequencing platforms.

However, only short sequences of DNA are determined from each DNA nanoball which makes mapping the short reads to a reference genome difficult. This technology has been used for multiple genome sequencing projects and is scheduled to be used for more.

Helioscope(TM) Single Molecule Sequencing

Based on "true single molecule sequencing" technology, Helioscope sequencing uses DNA fragments with added polyA tail adapters, which are attached to the flow cell surface. The next steps involve extension-based sequencing with cyclic washes of the flow cell with fluorescently labeled nucleotides (one nucleotide type at a time, as with the Sanger method).

The reads are performed by the Helioscope sequencer. The reads are short, up to 55 bases per run, but recent improvemend of the methodology allowes more accurate reads of homopolymers (stretches of one type of nucleotides) and RNA sequencing.

Single Molecule SMRT(TM) Sequencing

SMRT sequencing is based on the sequencing by synthesis approach. The DNA is synthesisd in so calles zero-mode wave-guides (ZMWs) - small well-like containers with the capturing tools located at the bottom of the well. The sequencing is performed with use of unmodified polymerase (attached to the ZMW bottom) and fluorescently labelled nucleotides flowing freely in the solution.

The wells are constructed in a way that only the fluorescence occurring by the bottom of the well is detected. The fluorescent label is detached from the nucleotide at its incorporation into the DNA strand, leaving an unmodified DNA strand. According to Pacific

Biosciences, the SMTR technology developer, this methodology allows detection of nucleotide modifications (such ad cytosine methylation). This happens through the observation of polymerase kinetics. This approach allows reads of 1000 nucleotides.

Single Molecule Real Time (RNAP) Sequencing

This method is based on RNA polymerase (RNAP), which is attached to a polystyrene bead, with distal end of sequenced DNA is attached to another bead, with both beads being placed in optical traps. RNAP motion during transcription brings the beads in closer and their relative distance changes, which can then be recorded at a single nucleotide resolution. The sequence is deduced based on the four readouts with lowered concentrations of each of the four nucleotide types.

Nanopore DNA Sequencing

This method is based on the readout of electrical signal occurring at nucleotides passing by alpha-hemolysin pores covalently bound with cyclodextrin. The DNA passing through the nanopore changes its ion current. This change is dependent on the shape, size and length of the DNA sequence.

Each type of the nucleotide blocks the ion flow through the pore for a different period of time. The method has a potential of development as it does not require modified nucleotides, however single nucleotide resolution is not yet available.

VisiGen Biotechnologies Approach

VisiGen Biotechnologies introduced a specially engineered DNA polymerase for use in their sequencing. This polymerase acts as a sensor - having incorporated a donor fluorescent dye by its active centre. This donor dye acts by FRET (fluorescent resonant energy transfer), inducing fluorescence of differently labeled nucleotides. This approach allows reads performed at the spead at which polymerase incorporates nucleotides into the sequence (several hundred per second). The nucleotide fluorochrome is released after the incorporation into the DNA strand. The expected read lengths in this approach should reach 1000 nucleotides, however this will have to be confirmed.

Future Methods

Sequencing by hybridization is a non-enzymatic method that uses a DNA microarray. A single pool of DNA whose sequence is to be determined is fluorescently labeled and hybridized to an array containing known sequences.

Strong hybridization signals from a given spot on the array identifies its sequence in the DNA being sequenced. Mass spectrometry may be used to determine mass differences between DNA fragments produced in chain-termination reactions.

DNA sequencing methods currently under development include labeling the DNA polymerase, reading the sequence as a DNA strand transits through nanopores, and microscopy-based techniques, such as AFM or transmission electron microscopy that are used to identify the positions of individual nucleotides within long DNA fragments (>5,000 bp) by nucleotide labeling with heavier elements (e.g., halogens) for visual detection and recording.

Third generation technologies aim to increase throughput and decrease the time to result and cost by eliminating the need for excessive reagents and harnessing the processivity of DNA polymerase.

In microfluidic Sanger sequencing the entire thermocycling amplification of DNA fragments as well as their separation by electrophoresis is done on a single glass wafer (approximately 10 cm in diameter) thus reducing the reagent usage as well as cost. In some instances researchers have shown that they can increase the throughput of conventional sequencing through the use of microchips. Research will still need to be done in order to make this use of technology effective.

In October 2006, the X Prize Foundation established an initiative to promote the development of full genome sequencing technologies, called the Archon X Prize, intending to award $10 million to "the first Team that can build a device and use it to sequence 100 human genomes within 10 days or less, with an accuracy of no more than one error in every 100,000 bases sequenced, with sequences accurately covering at least 98% of the genome, and at a recurring cost of no more than $10,000 (US) per genome."

Each year NHGRI promotes grants for new research and developments in genomics. 2010 grants and 2011 candidates include continuing work in microfluidic, polony and base-heavy sequencing methodologies

Major Landmarks in DNA Sequencing

- 1953 Discovery of the structure of the DNA double helix.

- 1972 Development of recombinant DNA technology, which permits isolation of defined fragments of DNA; prior to this, the only accessible samples for sequencing were from bacteriophage or virus DNA.
- 1977 The first complete DNA genome to be sequenced is that of bacteriophage öX174.
- 1977 Allan Maxam and Walter Gilbert publish "DNA sequencing by chemical degradation". Frederick Sanger, independently, publishes "DNA sequencing with chain-terminating inhibitors".
- 1984 Medical Research Council scientists decipher the complete DNA sequence of the Epstein-Barr virus, 170 kb.
- 1986 Leroy E. Hood's laboratory at the California Institute of Technology and Smith announce the first semi-automated DNA sequencing machine.
- 1987 Applied Biosystems markets first automated sequencing machine, the model ABI 370.
- 1990 The U.S. National Institutes of Health (NIH) begins large-
scale sequencing trials on Mycoplasma capricolum, Escherichia coli, Caenorhabditis elegans, and Saccharomyces cerevisiae (at US$0.75/base).
- 1991 Sequencing of human expressed sequence tags begins in Craig Venter's lab, an attempt to capture the coding fraction of the human genome.
- 1995 Craig Venter, Hamilton Smith, and colleagues at The Institute for Genomic Research (TIGR) publish the first complete genome of a free-living organism, the bacterium Haemophilus influenzae. The circular chromosome contains 1,830,137 bases and its publication in the journal Science marks the first use of whole-genome shotgun sequencing, eliminating the need for initial mapping efforts.
- 1996 Pål Nyrén and his student Mostafa Ronaghi at the Royal Institute of Technology in Stockholm publish their method of pyrosequencing
- 1998 Phil Green and Brent Ewing of the University of Washington publish "phred" for sequencer data analysis.

- 2000 Lynx Therapeutics publishes and markets "MPSS" - a parallelized, adapter/ligation-mediated, bead-based sequencing technology, launching "next-generation" sequencing.
- 2001 A draft sequence of the human genome is published.
- 2004 454 Life Sciences markets a parallelized version of pyrosequencing. The first version of their machine reduced sequencing costs 6-fold compared to automated Sanger sequencing, and was the second of a new generation of sequencing technologies, after MPSS.

Complete Genomics

Complete Genomics is a life sciences company that has developed and commercialized a DNA sequencing platform for human genome sequencing and analysis. This solution combines the company's proprietary human genome sequencing technology with its informatics and data management software in an outsourced service model that provides its customers with finished variant reports and assemblies. Researchers send DNA samples to the company, which are sequenced at Complete Genomics' own commercial genome center in Mountain View, California. Sequencing is completed using the company's proprietary instruments and technology. Assembled sequences and variant reports are then delivered back to the researcher.

History

Complete Genomics was founded in March 2006 by Clifford Reid, Radoje (Rade) Drmanac, and John Curson. Clifford Reid is the chairman, president and chief executive officer of Complete Genomics.

In Feb. 2009, Complete Genomics announced that it had sequenced its first human genome and submitted the resulting variant data to the National Center for Biotechnology Information database. Then in Nov. 2009, Complete Genomics published sequence data for three human genomes in the journal Science.

By the end of that year, Complete Genomics had sequenced 50 human genomes. The company has now significantly increased the throughput in its genome sequencing factory; it sequenced and analyzed 300 complete human genomes in Q3 2010.

In addition, Complete Genomics' customers are now publishing papers describing research completed using data from Complete Genomics. Examples include the Institute for Systems Biology's project to sequence a family of four and verify the gene responsible for Miller Syndrome, a rare craniofacial disorder and Genentech's work to sequence and compare a patient's primary lung tumor and adjacent normal tissue.

Technology Platform

Complete Genomics' proprietary human genome sequencing technology is optimized for the exclusive study of human DNA, providing assembled sequences and variation files. The technology relies onDNA nanoball sequencing. It is designed to use lower volumes and concentrations of reagents than existing systems. Additionally, Complete Genomics' proprietary instrumentation allows for greater numbers of base reads per image.

DNA Field-effect Transistor

A DNA field-effect transistor (DNAFET) is a field-effect transistor which uses the field-effect due to the partial charges of DNA molecules to function as a biosensor. The structure of DNAFETs is similar to that of MOSFETs with the exception of the gate structure which, in DNAFETs, is replaced by a layer of immobilized ssDNA (single-stranded DNA) molecules which act as surface receptors. When complementary DNA strands hybridize to the receptors, the charge distribution near the surface changes, which in turn modulates current transport through the semiconductor transducer.

Arrays of DNAFETs can be used for detecting single nucleotide polymorphisms (causing many hereditary diseases) and for DNA sequencing. Their main advantage compared to optical detection methods in common use today is that they do not require labeling of molecules.

DNA Sequencing Theory

DNA sequencing theory is the broad body of work that attempts to lay analytical foundations for DNA sequencing. The practical aspects revolve around designing and optimizing sequencing projects (known as "strategic genomics"), predicting project performance, troubleshooting experimental results, characterizing factors such as sequence bias and the effects of software processing algorithms, and comparing various sequencing methods to one another. In this sense, it could be considered a branch of systems engineering or operations research.

The permanent archive of work is primarily mathematical, although numerical calculations are often conducted for particular problems too. DNA sequencing theory addresses physical processes related to sequencing DNA and should not be confused with theories of analyzing resultant DNA sequences, e.g. sequence alignment. Publications sometimes do not make a careful distinction, but the latter are primarily concerned withalgorithmic issues.

Sequencing as a Covering Problem

All mainstream methods of DNA sequencing rely on reading small fragments of DNA and subsequently reconstructing these data to infer the original DNA target, either via assembly or alignment to a reference. The abstraction common to these methods is that of a mathematical covering problem.

For example, one can imagine a line segment representing the target and a subsequent process where smaller segments are "dropped" onto random locations of the target. The target is considered "sequenced" when adequate coverage accumulates, for example when no gaps remain.

The abstract properties of covering have been studied by mathematicians for over a century. However, direct application of these results has not generally been possible. Closed-form mathematical solutions, especially for probability distributions, often cannot be readily evaluated.

That is, they involve inordinately large amounts of computer time for parameters characteristic of DNA sequencing. Stevens' configuration is one such example. Results obtained from the perspective of pure mathematics also do not account for factors that are actually important in sequencing, for instance detectable overlap in sequencing fragments, double-stranding, edge-effects, and target multiplicity.

Consequently, development of sequencing theory has proceeded more according to the philosophy ofapplied mathematics. In particular, it has been problem-focused and makes expedient use of approximations, simulations, etc.

Early Uses Derived from Elementary Probability Theory

The earliest result was actually borrowed directly from elementary probability theory. If we model the above process and take L and G as the fragment length and target length, respectively, then the probability of "covering" any given location on the target with one particular fragment is L / G.

Note that this presumes $L \ll G$, which is valid for many, though not all sequencing scenarios. Utilizing concepts from the binomial distribution, it can then be shown that the probability that the location is covered by at least one of N fragments is

$$P = 1 - \left[1 - \frac{L}{G}\right]^N .$$

This equation was first used to characterize plasmid libraries, but is often more useful in a modified form. For most projects , so that, to a good degree of approximation

$$\left[1 - \frac{L}{G}\right]^N \sim \exp(-NL/G),$$

where R = NL / G is called the redundancy. Note the significance of redundancy as representing the average number of times a position is covered with fragments. Note also that in considering the covering process over all positions in the target, this probability is identical to the expected value of the random variable C, which represents the fraction of the target coverage. The final result,

$$E\langle C \rangle = 1 - e^{-R},$$

remains in widespread use as a "back of the envelope" estimator and predicts that coverage for all projects evolves along a universal curve that is a function only of the redundancy.

Lander-Waterman Theory

In 1988, Eric Lander and Michael Waterman published an important paper examining the covering problem from the standpoint of gaps. Although they focused on the so-called mapping problem, the abstraction to sequencing is much the same.

They furnished a number of useful results that were adopted as the standard theory from the earliest days of "large-scale" genome sequencing. Their model was also used in designing the Human Genome Project and continues to play an important role in DNA sequencing.

Ultimately, the main goal of a sequencing project is to close all gaps, so the "gap perspective" was a logical basis of developing a sequencing model. One of the more frequently used results from this model is the expected number of contigs, given the number of fragments sequenced. If one neglects the amount of sequence that is essentially "wasted" by having to detect overlaps, their theory yields

$$E\langle contings \rangle = Ne^{-R}.$$

In 1995, Roach published improvements to this theory, enabling it to be applied to sequencing projects in which the goal was to completely sequence a target genome. Wendl and Waterstonconfirmed, based on Stevens' method, that both models produced similar results when the number of contigs was substantial, such as in low coverage mapping or sequencing projects.

As sequencing projects ramped up in the 1990s, and projects approached completion, low coverage approximations became inadequate, and the exact model of Roach was necessary. However, as the cost of sequencing dropped, parameters of sequencing projects became easier to directly test empirically, and interest and funding for strategic genomics diminished

The basic ideas of Lander-Waterman theory led to a number of additional results for particular variations in mapping techniques. However, technological advancements have rendered mapping theories largely obsolete except in organisms other than highly studied model organisms (e.g., yeast, flies, mice, and humans).

Parking Strategy

The parking strategy for sequencing resembles the process of parking cars along a curb. Each car is a sequenced clone, and the curb is the genomic target. Each clone sequenced is screened to ensure that subsequently sequenced clones do not overlap any previously sequenced clone. No sequencing effort is redundant in this strategy. However, much like the gaps between parked cars, unsequenced gaps less than the length of a clone accumulate between sequenced clones. There can be considerable cost to close such gaps.

Pairwise End-sequencing

In 1995, Roach et al. proposed and demonstrated through simulations a generalization of a set of strategies explored earlier by Edwards and Caskey. This whole-genome sequencing method became immensely popular as it was championed by Celera and used to sequenced several model organisms before Celera applied it to the human genome. Today, most sequencing projects employ this strategy, often called paired end sequencing.

Recent Advancements

The physical processes and protocols of DNA sequencing have continued to evolve, largely driven by advancements in bio-chemical methods, hardware, and automation techniques. There is now a wide range of problems that DNA sequencing has made in-roads into, including metagenomics and medical (cancer) sequencing. There are important factors in these scenarios that classical theory does not account for. Recent work has begun to focus on resolving the effects of some of these issues. The level of mathematics becomes commensurately more sophisticated.

Multiplicity

Biologists have developed methods to filter highly-repetitive, essentially un-sequenceable regions of genomes. These procedures are important for organisms whose genomes consist mostly of such DNA, for example corn. They yield multitudes of small islands of sequenceable DNA products. Wendl and Barbazuk proposed an extension to Lander-Waterman Theory to account for "gaps" in the target due to filtering and the so-called "edge-effect".

The latter is a position-specific sampling bias, for example the terminal base position has only a 1 / G chance of being covered, as opposed toL / G for interior positions. For R < 1, classical Lander-Waterman Theory still gives good predictions, but dynamics change for higher redundancies.

Small Versus Large Fragments

Modern sequencing methods usually sequence both ends of a larger fragment, which provides linking information for de novo assembly and improved probabilities for alignment to reference sequence. Researchers generally believe that longer lengths of data (read lengths) enhance performance for very large DNA targets, an idea consistent with predictions from distribution models.

However, Wendl showed that smaller fragments provide better coverage on small, linear targets because they reduce the edge effect in linear molecules. These findings have implications for sequencing the products of DNA filtering procedures. Read-pairing and fragment size evidently have negligible influence for large, whole-genome class targets.

Diploid Sequencing

Sequencing is emerging as an important tool in medicine, for example in cancer research. Here, the ability to detect heterozygous mutations is important and this can only be done if the sequence of the diploid genome is obtained. In the pioneering efforts to sequence individuals, Levy et al. and Wheeler et al., who sequenced Craig Venter and Jim Watson, respectively, outlined models for covering both alleles in a genome.

Wendl and Wilson followed with a more general theory that allowed for an arbitrary number of coverings of each allele and arbitrary ploidy. These results point to the general conclusion that the amount of data needed for such projects is significantly higher than for traditional haploid projects.

Limitations

DNA sequencing theories often invoke the assumption that certain random variables in a model are independently and identically distributed. For example, in Lander-Waterman Theory, a sequenced fragment is presumed to have the same probability of covering each region of a genome and all fragments are assumed to be independent of one another.

In actuality, sequencing projects are subject to various types of bias, including differences of how well regions can be cloned, sequencing anomalies, biases in the target sequence (which is not random), and software-dependent errors and biases. In general, theory will agree well with observation up to the point that enough data have been generated to expose latent biases. The kinds of biases related to the underlying target sequence are particularly difficult to model, since the sequence itself may not be known a priori. This presents a type of "chicken and egg" closure problem.

Academic Status

Sequencing theory is based on elements of mathematics, biology, and systems engineering, so it is highly interdisciplinary. Although many universities now have programs in computational biology, there does not yet seem to be a strong focus at the graduate level on this topic. Academic contributions have mainly been limited to a small number of PhD dissertations.

Chapter 3

Genetic Assembly

Genome assembly refers to the process of taking a large number of short DNA sequences and putting them back together to create a representation of the original chromosomes from which the DNA originated. In a shotgun sequencing project, all the DNA from a source (usually a single organism, anything from a bacterium to a mammal) is first fractured into millions of small pieces. These pieces are then "read" by automated sequencing machines, which can read up to 900 nucleotides or bases at a time.

(The four bases are adenine, guanine, cytosine, and thymine, represented as AGCT.) A genome assembly algorithm works by taking all the pieces and aligning them to one another, and detecting all places where two of the short sequences, or reads, overlap. These overlapping reads can be merged together, and the process continues.

Genome assembly is a very difficult computational problem, made more difficult because many genomes contain large numbers of identical sequences, known as repeats.

These repeats can be thousands of nucleotides long, and some occur in thousands of different locations, especially in the large genomes of plants and animals.

The resulting (draft) genome sequence is produced by combining the information sequenced contigs and then employing linking information to create scaffolds. Scaffolds are positioned along thephysical map of the chromosomes creating a "golden path".

Assembly Software

Originally, most large-scale DNA sequencing centers developed their own software for assembling the sequences that they produced. However, this has changed as the software has grown more complex and as the number of sequencing centers has increased. An example of such assembler Short Oligonucleotide Analysis Package developed by BGI for de novo assembly of human-sized genomes, alignment, SNP detection, resequencing, indel finding, and structural variation analysis.

Genome Annotation

Genome annotation is the process of attaching biological information to sequences. It consists of two main steps:

1. identifying elements on the genome, a process called gene prediction, and
2. attaching biological information to these elements.

Automatic annotation tools try to perform all this by computer analysis, as opposed to manual annotation (a.k.a. curation) which involves human expertise. Ideally, these approaches co-exist and complement each other in the same annotation pipeline.

The basic level of annotation is using BLAST for finding similarities, and then annotating genomes based on that. However, nowadays more and more additional information is added to the annotation platform.

The additional information allows manual annotators to deconvolute discrepancies between genes that are given the same annotation.

Some databases use genome context information, similarity scores, experimental data, and integrations of other resources to provide genome annotations through their Subsystems approach. Other databases (e.g. Ensembl) rely on both curated data sources as well as a range of different software tools in their automated genome annotation pipeline.

Structural annotation consists of the identification of genomic elements.

- ORFs and their localisation
- gene structure
- coding regions
- location of regulatory motifs

Functional annotation consists of attaching biological information to genomic elements.

- biochemical function

- biological function
- involved regulation and interactions
- expression

These steps may involve both biological experiments and in silico analysis.

A variety of software tools have been developed to permit scientists to view and share genome annotations.

Genome annotation is the next major challenge for the Human Genome Project, now that the genome sequences of human and several model organisms are largely complete.

Identifying the locations of genes and other genetic control elements is often described as defining the biological "parts list" for the assembly and normal operation of an organism. Scientists are still at an early stage in the process of delineating this parts list and in understanding how all the parts "fit together".

Genome annotation is an active area of investigation and involves a number of different organizations in the life science community which publish the results of their efforts in publicly available biological databases accessible via the web and other electronic means. Here is an alphabetical listing of on-going projects relevant to genome annotation:

- ENCyclopedia Of DNA Elements (ENCODE)
- Entrez Gene
- Ensembl
- GENCODE
- Gene Ontology Consortium
- GeneRIF
- RefSeq
- Uniprot
- Vertebrate and Genome Annotation Project (Vega)

At Wikipedia, genome annotation has started to become automated under the auspices of the Gene Wiki portal which operates a bot that harvests gene data from research databases and creates gene stubs on that basis.

Genome Project Finished

A genome, there are usually regions that are difficult to sequence (often regions with highly repetitive DNA). Thus, 'completed' genome sequences are rarely ever complete, and terms such as 'working draft' or 'essentially complete' have been used to more accurately describe the status of such genome projects.

Even when every base pair of a genome sequence has been determined, there are still likely to be errors present because DNA sequencing is not a completely accurate process. It could also be argued that a complete genome project should include the sequences of mitochondria and (for plants) chloroplasts as these organelles have their own genomes.

It is often reported that the goal of sequencing a genome is to obtain information about the complete set of genes in that particular genome sequence. The proportion of a genome that encodes for genes may be very small (particularly in eukaryotes such as humans, where coding DNA may only account for a few percent of the entire sequence).

However, it is not always possible (or desirable) to only sequence the coding regions separately. Also, as scientists understand more about the role of this noncoding DNA (often referred to as junk DNA), it will become more important to have a complete genome sequence as a background to understanding the genetics and biology of any given organism.

In many ways genome projects do not confine themselves to only determining a DNA sequence of an organism. Such projects may also include gene prediction to find out where the genes are in a genome, and what those genes do. There may also be related projects to sequence ESTs or mRNAs to help find out where the genes actually are.

Historical and Technological Perspectives

Historically, when sequencing eukaryotic genomes (such as the worm Caenorhabditis elegans) it was common to first map the genome to provide a series of landmarks across the genome. Rather than sequence a chromosome in one go, it would be sequenced piece by piece (with the prior knowledge of approximately where that piece is located on the larger chromosome).

Changes in technology and in particular improvements to the processing power of computers, means that genomes can now be 'shotgun sequenced' in one go (there are caveats to this approach though when compared to the traditional approach).

Improvements in DNA sequencing technology has meant that the cost of sequencing a new genome sequence has steadily fallen (in terms of cost per base pair) and newer technology has also meant that genomes can be sequenced far more quickly.

When research agencies decide what new genomes to sequence, the emphasis has been on species which are either high importance as model organism or have a relevance to human health (e.g. pathogenic bacteria or vectors of disease such as mosquitos) or species which have commercial importance (e.g. livestock and crop plants). Secondary emphasis is placed on species whose genomes will help answer important questions in molecular evolution (e.g. the common chimpanzee).

In the future, it is likely that it will become even cheaper and quicker to sequence a genome. This will allow for complete genome sequences to be determined from many different individuals of the same species. For humans, this will allow us to better understand aspects of human genetic diversity.

Genome Projects

Fig. 1: L1 Dominette 01449, the Hereford who serves as the subject of the Bovine Genome Project

Many organisms have genome projects that have either been completed or will be completed shortly, including:

- Humans, Homo sapiens;
- Palaeo-Eskimo, an ancient-human

- Neanderthal, "Homo neanderthalensis".
- Common Chimpanzee Pan troglodytes; s
- Domestic Cow
- Bovine Genome
- Honey Bee Genome Sequencing Consortium
- Human microbiome project
- International Grape Genome Program
- International HapMap Project

Multiplex Ligation-dependent Probe Amplification

Multiplex ligation-dependent probe amplification (MLPA) is a variation of the polymerase chain reaction that permits multiple targets to be amplified with only a single primer pair.

Each probe consists of a two oligonucleotides which recognise adjacent target sites on the DNA. One probe oligonucleotide contains the sequence recognised by the forward primer, the other the sequence recognised by the reverse primer. Only when both probe oligonucleotides are hybridised to their respective targets, can they be ligated into a complete probe.

The advantage of splitting the probe into two parts is that only the ligated oligonucleotides, but not the unbound probe oligonucleotides, are amplified. If the probes were not split in this way, the primer sequences at either end would cause the probes to be amplified regardless of their hybridization to the template DNA, and the amplification product would not be dependent on the number of target sites present in the sample DNA.

Each complete probe has a unique length, so that its resulting amplicons can be separated and identified by (capillary) electrophoresis.

This avoids the resolution limitations of multiplex PCR. Since the forward primer used for probe amplification is fluorescently labeled, each amplicon generates a fluorescent peak which can be detected by a capillary sequencer.

Comparing the peak pattern obtained on a given sample with that obtained on various reference samples, the relative quantity of each amplicon can be determined. This ratio is a measure for the ratio in which the target sequence is present in the sample DNA.

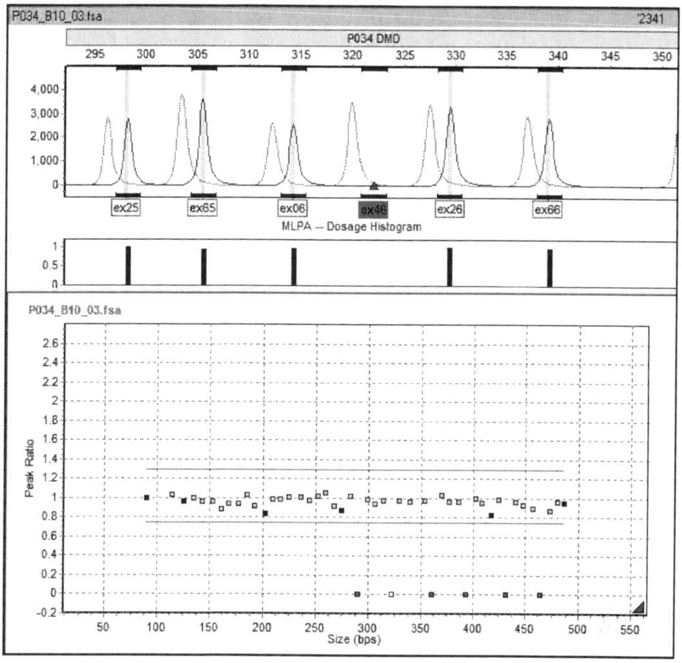

Fig. 2:

Various techniques including DGGE (Denaturing Gradient Gel Electrophoresis), DHPLC (Denaturing High Performance Liquid Chromatography), and SSCA (Single Strand Conformation Analysis) effectively identify SNPs and small insertions and deletions. MLPA, however, is one of the only accurate, time-efficient techniques to detect genomic deletions and insertions (one or more entire exons), which are frequent causes of cancers such as hereditary non-polyposis colorectal cancer (HNPCC), breast, and ovarian cancer. MLPA can successfully and easily determine the relative copy number of all exons within a gene simultaneously with high sensitivity.

Relative Ploidy

An important use of MLPA is to determine relative ploidy. For example, probes may be designed to target various regions of chromosome 21 of a human cell.

The signal strengths of the probes are compared with those obtained from a reference DNA sample known to have two copies of the chromosome. If an extra copy is present in the test sample, the signals are expected to be 1.5 times the intensities of the respective probes from the reference. If only one copy is present the proportion is expected to be 0.5. If the sample has two copies, the relative probe strengths are expected to be equal.

Dosage Quotient Analysis

Dosage quotient analysis is the usual method of interpreting MLPA data. If a and b are the signals from two amplicons in the patient sample, and A and B are the corresponding amplicons in the experimental control, then the dosage quotient DQ = (a/b) / (A/B). Although dosage quotients may be calculated for any pair of amplicons, it is usually the case that one of the pair is an internal reference probe.

Advantages of MLPA

MLPA facilitates the amplification and detection of multiple targets with a single primer pair. In a standard multiplex PCR reaction, each fragment needs a unique amplifying primer pair. These primers being present in a large quantity result in various problems such as dimerization and false priming. With MLPA, amplification of probes can be achieved. Thus, many sequences (up to 40) can be amplified and quantified using just a single primer pair. MLPA reaction is fast, cheap and very simple to perform.

Applications of MLPA

MLPA has a variety of applications including detection of mutations and single nucleotide polymorphisms, analysis of DNA methylation, relative mRNA quantification, chromosomal characterisation of cell lines and tissue samples, detection of gene copy number,detection of duplications and deletions in human cancer predisposition genes such as BRCA1, BRCA2, hMLH1and hMSH2 and aneuploidy determination. MLPA has potential application in prenatal diagnosis both invasive and noninvasive.

Sequence Profiling tool

A sequence profiling tool in bioinformatics is a type of software that presents information related to a genetic sequence, gene name, or keyword input.

Such tools generally take a query such as aDNA, RNA, or protein sequence or 'keyword' and search one or more databases for information related to that sequence. Summaries and aggregate results are provided in standardized format describing the information that would otherwise have required visits to many smaller sites or direct literature searches to compile.

Many sequence profiling tools are software portals or gateways that simplify the process of finding information about a query in the large and growing number of bioinformatics databases. The access to these kinds of tools is either web based or locally downloadable executables.

Introduction and Usage

The "post-genomics" era has given rise to a range of web-based tools and software to compile, organize, and deliver large amounts of primary sequence information, as well as protein structures, gene annotations, sequence alignments, and other common bioinformatics tasks.

In general, there exist three types of databases and service providers. The first one includes the popular public-domain or open-access databases supported by funding and grants such as NCBI,ExPASy, Ensembl, and PDB. The second one includes smaller or more specific databases organized and compiled by individual research groups Examples include Yeast Genome Database, RNA database.

The third and final one includes private corporate or institutional databases that require payment or institutional affiliation to access. Such examples rare given the globalization of the public databases unless the purported service is 'in-development' or the end point of the analysis is of commercial value.

Typical scenarios of a profiling approach become relevant, particularly, in the cases of the first two groups, where researchers commonly wish to combine information derived from several sources about a single query or target sequence.

For example, users might use the sequence alignment and search tool BLAST to identify homologs of their gene of interest in other species, and then use these results to locate a solved protein structure for one of the homologs.

Similarly, they might also want to know the likely secondary structure of the mRNA encoding the gene of interest, or whether a company sells a DNA construct containing the gene. Sequence profiling tools serve to automate and integrate the process of seeking such disparate information by rendering the process of searching several different external databases transparent to the user.

Many public databases are already extensively linked so that complementary information in another database is easily accessible; for example, Genbank and the PDB are closely intertwined. However, specialized tools organized and hosted by specific research groups can be difficult to integrate into this linkage effort because they are narrowly focused, are frequently modified, or use custom versions of common file formats.

Advantages of sequence profiling tools include the ability to use multiple of these specialized tools in a single query and present the output with a common interface, the ability to direct the output of one set of tools or database searches into the input of another, and the capacity to disseminate hosting and compilation obligations to a network of research groups and institutions rather than a single centralized repository.

Keyword Based Profilers

Most of the profiling tools available on the web today fall into this category. The user, upon visiting the site/tool, enters any relevant information like a keyword e.g. dystrophy, diabetes etc., or GenBank accession numbers, PDB ID.

All the relevant hits by the search are presented in a format unique to each tool's main focus. Profiling tools based on keyword searches are essentially search enginesthat are highly specialized for bioinformatics work, thereby eliminating a clutter of irrelevant or non-scholarly hits that might occur with a traditional search engine like Google. Most keyword-based profiling tools allow flexible types of keyword input, accession numbers from indexed databases as well as traditional keyword descriptors.

Each profiling tool has its own focus and area of interest. For example, the NCBI search engine Entrez segregates its hits by category, so that users looking for protein structure information can screen out sequences with no corresponding structure, while users interested in perusing the literature on a subject can view abstracts of papers published in scholarly journals without distraction from gene or sequence results. The Pubmed biosciences literature database is a popular tool for literature searches, though this service is nearly equaled with the more general Google Scholar.

Keyword-based data aggregation services like the Bioinformatic Harvester performs provide reports from a variety of third-party servers in an as-is format so that users need not visit the website or install the software for each individual component service.

This is particularly invaluable given the rapid emergence of various sites providing different sequence analysis and manipulation tools. Another aggregative web portal, the Human Protein Reference Database (Hprd), contains manually annotated and curated entries for human proteins.

The information provided is thus both selective and comprehensive, and the query format is flexible and intuitive. The pros of developing manually curated databases include presentation of proofread material and the concept of 'molecule authorities' to undertake the responsibility of specific proteins. However, the cons are that they are typically slower to update and may not contain very new or disputed data.

Sequence Data Based Profilers

A typical sequence profiling tool carries this further by using an actual DNA, RNA, or protein sequence as an input and allows the user to visit different web-based analysis tools to obtain the information desired. Such tools are also commonly supplied with commercial laboratory equipment like gene sequencers or sometimes sold as software applications for molecular biology.

In another public-database example, the BLAST sequence search report from NCBI provides a link from its alignment report to other relevant information in its own databases, if such specific information exists.

For example, a retrieved record that contains a human sequence will carry a separate link that connects to its location on a human genome map; a record that contains a sequence for which a 3-D structure has been solved would carry a link that connects it to its structure database. Sequerome, a public service tool, links the entire BLAST report to many third party servers/sites that provide highly specific services in sequence manipulations such as restriction enzyme maps, open reading frame analyses for nucleotide sequences, and secondary structure prediction.

The tool provides added advantage of maintaining a research log of the operations performed by the user, which can be then conveniently archived using 'mail', 'print' or 'save' functionality. Thus an entire operation of researching on a sequence using different research tools and thus carrying a project to its completion can be completed within one browser interface.

Consequently, future generation of sequence profiling tools would include ability to collaborate online with researchers to share project logs and research tools, annotate results of sequence analysis or lab work, customize and automate the processing of sets of sequence data etc.

InstaSeq is a Google powered search tool that allows the user to directly enter a sequence and search the entire World Wide Web. This unique search engine, which is the only one of its kind, is in contrast to searching specific databases e.g.GenBank.

As a result the user can end up with a privately hosted document or a page from a lesser known database from just about anywhere in the world. Though the presence of sequence based profilers are far and few in the present scenario, their key role will become evident when huge amounts of sequence data need to be cross processed across portals and domains.

Future Growth and Directions

The proliferation of bioinformatics tools for genetic analysis aids researchers in identifying and categorizing genes and gene sets of interest in their work; however, the large variety of tools that perform substantially similar aggregative and analytical functions can also confuse and frustrate new users.

The decentralization encouraged by aggregative tools allows individual research groups to maintain specialized servers dedicated to specific types of data analysis in the expectation that their output will be collected into a larger report on a gene or protein of interest to other researchers.

Data produced by microarray experiments, two-hybrid screening, and other high-throughput biological experiments is voluminous and difficult to analyze by hand; the efforts of structural genomic-scollaborations that are aimed at quickly solving large numbers of highly varied protein structures also increase the need for integration between sequence and structure databases and portals.

This impetus toward developing more comprehensive and more user-friendly methods of sequence profiling makes this an active area of research among current genomics researchers.

DNA Sequencing

DNA sequencing includes several methods and technologies that are used for determining the order of the nucleotide bases—adenine, guanine, cytosine, and thymine—in a molecule of DNA.

Knowledge of DNA sequences has become indispensable for basic biological research, other research branches utilizing DNA sequencing, and in numerous applied fields such as diagnostic, biotechnology, forensic biology and biological systematics. The advent of DNA sequencing has significantly accelerated biological research and discovery.

The rapid speed of sequencing attained with modern DNA sequencing technology has been instrumental in the sequencing of the human genome, in the Human Genome Project. Related projects, often by scientific collaboration across continents, have generated the complete DNA sequences of many animal, plant, and microbial genomes.

The first DNA sequences were obtained in the early 1970s by academic researchers using laborious methods based on two-dimensional chromatography. Following the development of dye-based sequencing methods with automated analysis, DNA sequencing has become easier and orders of magnitude faster.

History

RNA sequencing was one of the earliest forms of nucleotide sequencing. The major landmark of RNA sequencing is the sequence of the first complete gene and the complete genome of Bacteriophage MS2, identified and published by Walter Fiers and his coworkers at the University of Ghent (Ghent, Belgium), between 1972 and 1976.

Prior to the development of rapid DNA sequencing methods in the early 1970s by Frederick Sanger at the University of Cambridge, in England and Walter Gilbert and Allan Maxam at Harvard, a number of laborious methods were used. For instance, in 1973, Gilbert and Maxam reported the sequence of 24 basepairs using a method known as wandering-spot analysis.

The chain-termination method developed by Sanger and coworkers in 1977 soon became the method of choice, owing to its relative ease and reliability. It involves separating DNA bases from different DNA fragments.

Maxam–Gilbert Sequencing

In 1976–1977, Allan Maxam and Walter Gilbert developed a DNA sequencing method based on chemical modification of DNA and subsequent cleavage at specific bases. Although Maxam and Gilbert published their chemical sequencing method two years after the ground-breaking paper of Sanger and Coulson on plus-minus sequencing, Maxam–Gilbert sequencing rapidly became more popular, since purified DNA could be used directly, while the initial Sanger method required that each read start be cloned for production of single-stranded DNA.

However, with the improvement of the chain-termination method Maxam-Gilbert sequencing has fallen out of favour due to its technical complexity prohibiting its use in standard molecular biology kits, extensive use of hazardous chemicals, and difficulties with scale-up.

The method requires radioactive labeling at one 5' end of the DNA (typically by a kinase reaction using gamma-P ATP) and purification of the DNA fragment to be sequenced. Chemical treatment generates breaks at a small proportion of one or two of the four nucleotide bases in each of four reactions (G, A+G, C, C+T).

For example, the purines (A+G) are depurinated using formic acid, the guanines (and to some extent the adenines) are methylated by dimethyl sulfate, and the pyrimidines (C+T) are methylated using hydrazine. The addition of salt (sodium chloride) to the hydrazine reaction inhibits the methylation of thymine for the C-only reaction.

The modified DNAs are then cleaved by hot piperidine at the position of the modified base. The concentration of the modifying chemicals is controlled to introduce on average one modification per DNA molecule. Thus a series of labeled fragments is generated, from the radiolabeled end to the first "cut" site in each molecule. The fragments in the four reactions are electrophoresed side by side in denaturing acrylamide gels for size separation. To visualize the fragments, the gel is exposed to X-ray film for autoradiography, yielding a series of dark bands each corresponding to a radiolabeled DNA fragment, from which the sequence may be inferred.

Also sometimes known as "chemical sequencing", this method led to the Methylation Interference Assay used to map DNA-binding sites for DNA-binding proteins.

Chain-termination Methods

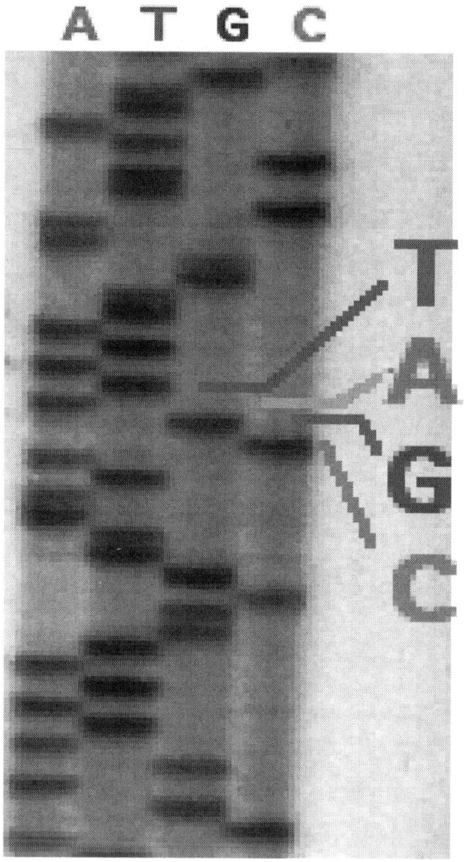

Fig. 3: Part of a radioactively labelled sequencing gel

Because the chain-terminator method (or Sanger method after its developer Frederick Sanger) is more efficient and uses fewer toxic chemicals and lower amounts of radioactivity than the method of Maxam and Gilbert, it rapidly became the method of choice. The key principle of the Sanger method was the use of dideoxy-nucleotidetriphosphates (ddNTPs) as DNA chain terminators.

The classical chain-termination method requires a single-stranded DNA template, a DNA primer, a DNA polymerase, normal deoxynucleotidetriphosphates (dNTPs), and modified nucleotides (dideoxyNTPs) that terminate DNA strand elongation. These ddNTPs will also be radioactively or fluorescently labelled for detection in automated sequencing machines. The DNA sample is divided into four separate sequencing reactions, containing all four of the standard deoxynucleotides (dATP, dGTP, dCTP and dTTP) and the DNA polymerase. To each reaction is added only one of the four dideoxynucleotides (ddATP, ddGTP, ddCTP, or ddTTP) which are the chain-terminating nucleotides, lacking a 3'-OH group required for the formation of a phosphodiester bond between two nucleotides, thus terminating DNA strand extension and resulting in DNA fragments of varying length.

The newly synthesized and labelled DNA fragments are heat denatured, and separated by size (with a resolution of just one nucleotide) by gel electrophoresis on a denaturing polyacrylamide-urea gel with each of the four reactions run in one of four individual lanes (lanes A, T, G, C); the DNA bands are then visualized byautoradiography or UV light, and the DNA sequence can be directly read off the X-ray film or gel image. In the image on the right, X-ray film was exposed to the gel, and the dark bands correspond to DNA fragments of different lengths. A dark band in a lane indicates a DNA fragment that is the result of chain termination after incorporation of a dideoxynucleotide (ddATP, ddGTP, ddCTP, or ddTTP). The relative positions of the different bands among the four lanes are then used to read (from bottom to top) the DNA sequence.

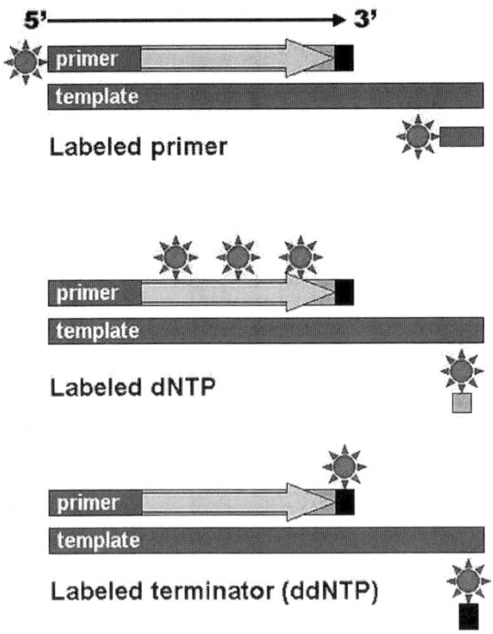

FIg. 4:

DNA fragments are labelled with a radioactive or fluorescent tag on the primer, in the new DNA strand with a labeled dNTP, or with a labeled ddNTP.

Technical variations of chain-termination sequencing include tagging with nucleotides containing radioactive phosphorus forradiolabelling, or using a primer labeled at the 5' end with a fluorescent dye. Dye-primer sequencing facilitates reading in an optical system for faster and more economical analysis and automation. The later development by Leroy Hood and coworkers of fluorescently labeled ddNTPs and primers set the stage for automated, high-throughput DNA sequencing.

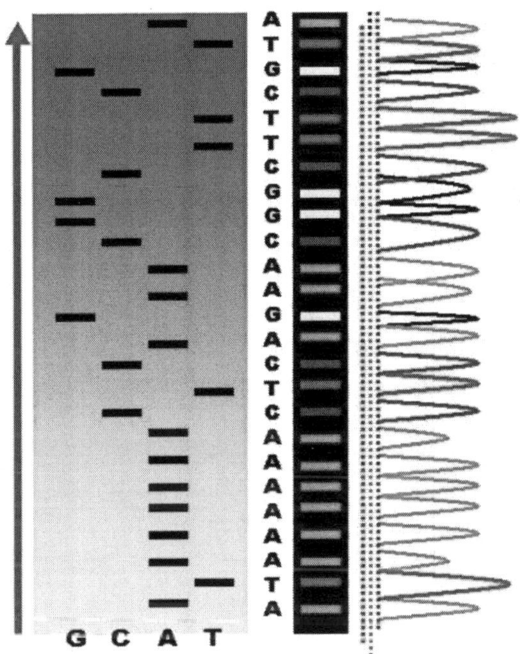

Fig. 5: Sequence ladder by radioactive sequencing compared to fluorescent peaks (click to expand)

Chain-termination methods have greatly simplified DNA sequencing. For example, chain-termination-based kits are commercially available that contain the reagents needed for sequencing, pre-aliquoted and ready to use. Limitations include non-specific binding of the primer to the DNA, affecting accurate read-out of the DNA sequence, and DNA secondary structures affecting the fidelity of the sequence.

Dye-terminator Sequencing

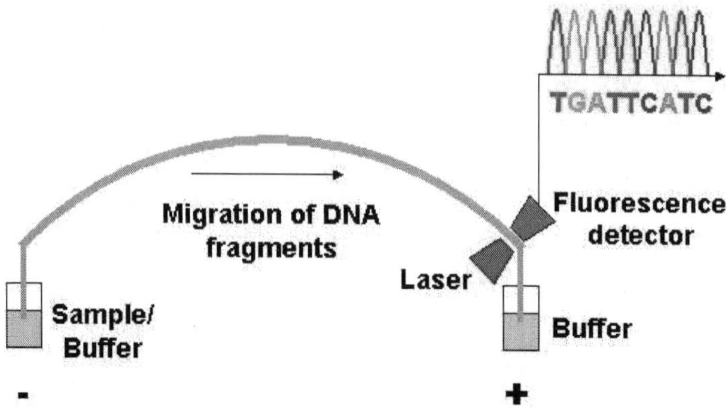

Fig. 6: Capillary electrophoresis (click to expand)

Dye-terminator sequencing utilizes labelling of the chain terminator ddNTPs, which permits sequencing in a single reaction, rather than four reactions as in the labelled-primer method. In dye-terminator sequencing, each of the four dideoxynucleotide chain terminators is labelled with fluorescent dyes, each of which emit light at differentwavelengths.

Owing to its greater expediency and speed, dye-terminator sequencing is now the mainstay in automated sequencing. Its limitations include dye effects due to differences in the incorporation of the dye-labelled chain terminators into the DNA fragment, resulting in unequal peak heights and shapes in the electronic DNA sequence trace chromatogram after capillary electrophoresis.

This problem has been addressed with the use of modified DNA polymerase enzyme systems and dyes that minimize incorporation variability, as well as methods for eliminating "dye blobs". The dye-terminator sequencing method, along with automated high-throughput DNA sequence analyzers, is now being used for the vast majority of sequencing projects.

Challenges

Common challenges of DNA sequencing include poor quality in the first 15–40 bases of the sequence and deteriorating quality of sequencing traces after 700–900 bases. Base calling software typically gives an estimate of quality to aid in quality trimming.

In cases where DNA fragments are cloned before sequencing, the resulting sequence may contain parts of the cloning vector. In contrast, PCR-based cloning and emerging sequencing technologies based on pyrosequencing often avoid using cloning vectors.

Recently, one-step Sanger sequencing (combined amplification and sequencing) methods such as Ampliseq and SeqSharp have been developed that allow rapid sequencing of target genes without cloning or prior amplification.

Current methods can directly sequence only relatively short (300–1000 nucleotides long) DNA fragments in a single reaction. The main obstacle to sequencing DNA fragments above this size limit is insufficient power of separation for resolving large DNA fragments that differ in length by only one nucleotide. In all cases the use of a primer with a free 3' end is essential.

Automation and Sample Preparation

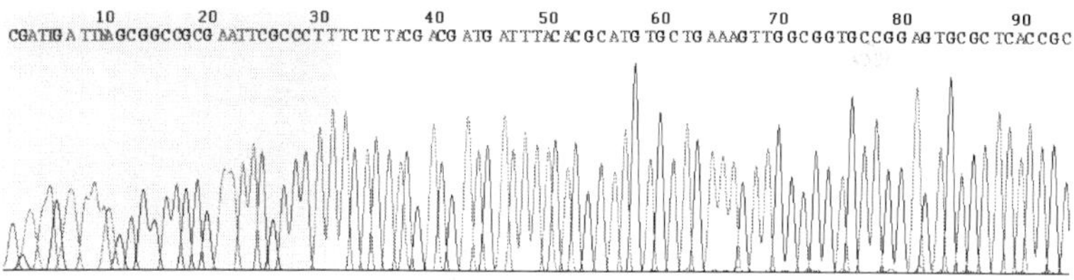

Fig. 7: View of the start of an example dye-terminator read (click to expand)

Automated DNA-sequencing instruments (DNA sequencers) can sequence up to 384 DNA samples in a single batch (run) in up to 24 runs a day.

DNA sequencers carry out capillary electrophoresis for size separation, detection and recording of dye fluorescence, and data output as fluorescent peak tracechromatograms. Sequencing reactions by thermocycling, cleanup and re-suspension in a buffer solution before loading onto the sequencer are performed separately. A number of commercial and non-commercial software packages can trim low-quality DNA traces automatically.

These programs score the quality of each peak and remove low-quality base peaks (generally located at the ends of the sequence). The accuracy of such algorithms is below visual examination by a human operator, but sufficient for automated processing of large sequence data sets.

Amplification and Clonal Selection

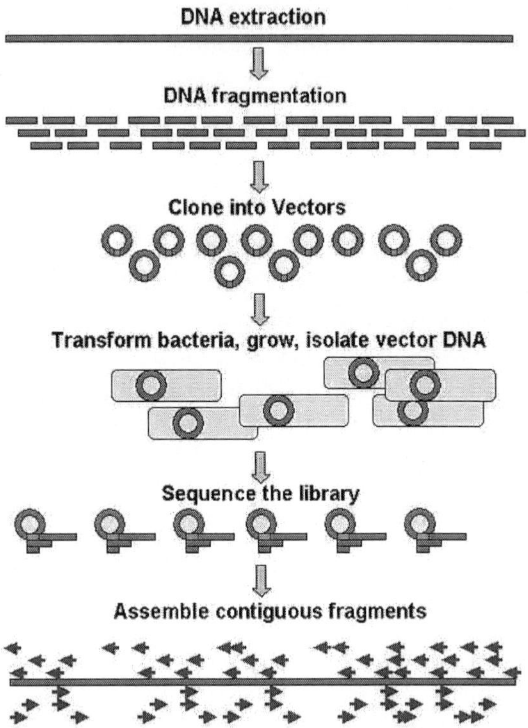

Fig. 8: Genomic DNA is fragmented into random pieces and cloned as a bacterial library. DNA from individual bacterial clones is sequenced and the sequence is assembled by using overlapping DNA regions.(click to expand)

Large-scale sequencing often aims at sequencing very long DNA pieces, such as whole chromosomes, although large scale sequencing can also be used to generate very large numbers of short sequences, such as found in phage display. For longer targets, such as chromosomes, common approaches consist of cutting (with restriction enzymes) or shearing (with mechanical forces) large DNA fragments into shorter DNA fragments. The fragmented DNA is cloned into aDNA vector, and amplified in Escherichia coli. Short DNA fragments purified from individual bacterial colonies are individually sequenced and assembled electronically into one long, contiguous sequence.

This method does not require any pre-existing information about the sequence of the DNA and is referred to as de novo sequencing. Gaps in the assembled sequence may be filled by primer walking. The different strategies have different tradeoffs in speed and accuracy;

shotgun methods are often used for sequencing large genomes, but its assembly is complex and difficult, particularly with sequence repeats often causing gaps in genome assembly.

Most sequencing approaches use an in vitro cloning step to amplify individual DNA molecules, because their molecular detection methods are not sensitive enough for single molecule sequencing. Emulsion PCR isolates individual DNA molecules along with primer-coated beads in aqueous droplets within an oil phase. Polymerase chain reaction (PCR) then coats each bead with clonal copies of the DNA molecule followed by immobilization for later sequencing. Emulsion PCR is used in the methods by Marguilis et al. (commercialized by 454 Life Sciences), Shendure and Porreca et al. (also known as "Polony sequencing") and SOLiD sequencing, (developed by Agencourt, now Applied Biosystems).

Another method for in vitro clonal amplification is bridge PCR, where fragments are amplified upon primers attached to a solid surface, used in the IlluminaGenome Analyzer. Single-molecule methods, such as that developed by Stephen Quake's laboratory (later commercialized by Helicos) is an exception: it uses bright fluorophores and laser excitation to detect base addition events from individual DNA molecules fixed to a surface, eliminating the need for molecular amplification.

High-throughput Sequencing

The high demand for low-cost sequencing has driven the development of high-throughput sequencing technologies that parallelize the sequencing process, producing thousands or millions of sequences at once. High-throughput sequencing technologies are intended to lower the cost of DNA sequencing beyond what is possible with standard dye-terminator methods.

The first of the "next-generation" sequencing technologies, MPSS was developed in the 1990s at Lynx Therapeutics, a company founded in 1992 by Sydney Brenner and Sam Eletr. MPSS was a bead-based method that used a complex approach of adapter ligation followed by adapter decoding, reading the sequence in increments of four nucleotides; this method made it susceptible to sequence-specific bias or loss of specific sequences. Because the technology was so complex, MPSS was only performed 'in-house' by Lynx Therapeutics and no machines were sold; when the merger with Solexa later led to the development of sequencing-by-synthesis, a more simple approach with numerous advantages, MPSS became obsolete. However, the essential properties of the MPSS output were typical of later "next-gen" data types, including hundreds of thousands of short DNA sequences. In the case of MPSS, these were typically used for sequencing cDNA for measurements of gene expression levels. Lynx Therapeutics merged with Solexa in 2004, and this company was later purchased by Illumina.

Polony Sequencing

Polony sequencing, developed in the laboratory of George Church at Harvard, was among the first next-generation sequencing systems used to sequence a full genome in 2005. It combined an in vitro paired-tag library with emulsion PCR, an automated microscope, and ligation-based sequencing chemistry to sequence an E. coli genome at an accuracy of > 99.9999% and a cost approximately 1/10 that of Sanger sequencing. The technology was licensed to Agencourt Biosciences, subsequently spun out into Agencourt Personal Genomics, and ultimately incorporated into the Applied Biosystems SOLiD platform.

Pyrosequencing

A parallelized version of pyrosequencing was developed by 454 Life Sciences, which has since been acquired by Roche Diagnostics. The method amplifies DNA inside water droplets in an oil solution (emulsion PCR), with each droplet containing a single DNA template attached to a single primer-coated bead that then forms a clonal colony. The sequencing machine contains many picolitre-volume wells each containing a single bead and sequencing enzymes. Pyrosequencing uses luciferase to generate light for detection of the individual nucleotides added to the nascent DNA, and the combined data are used to generate sequence read-outs. This technology provides intermediate read length and price per base compared to Sanger sequencing on one end and Solexa and SOLiD on the other.

Illumina (Solexa) Sequencing

Solexa, now part of Illumina, developed a sequencing technology based on reversible dye-terminators. DNA molecules are first attached to primers on a slide and amplified so that local clonal colonies are formed (bridge amplification). Four types of ddNTPs are added, and non-incorporated nucleotides are washed away. Unlike pyrosequencing, the DNA can only be extended one nucleotide at a time. A camera takes images of the fluorescently labeled nucleotides, then the dye along with the terminal 3' blocker is chemically removed from the DNA, allowing the next cycle.

SOLiD Sequencing

Applied Biosystems' SOLiD technology employs sequencing by ligation. Here, a pool of all possible oligonucleotides of a fixed length are labeled according to the sequenced position. Oligonucleotides are annealed and ligated; the preferential ligation by DNA ligase for matching sequences results in a signal informative of the nucleotide at that position. Before sequencing, the DNA is amplified by emulsion PCR. The resulting bead, each containing only copies of the same DNA molecule, are deposited on a glass slide. The result is sequences of quantities and lengths comparable to Illumina sequencing.

Ion Semiconductor Sequencing

Ion Torrent Systems Inc. developed a system based on using standard sequencing chemistry, but with a novel, semiconductor based detection system. This method of sequencing is based on the detection of hydrogen ions that are released during the polymerisation of DNA, as opposed to the optical methods used in other sequencing systems. A microwell containing a template DNA strand to be sequenced is flooded with a single type of nucleotide. If the introduced nucleotide is complementary to the leading template nucleotide it is incorporated into the growing complementary strand. This causes the release of a hydrogen ion that triggers a hypersensitive ion sensor, which indicates that a reaction has occurred. If homopolymer repeats are present in the template sequence multiple nucleotides will be incorporated in a single cycle. This leads to a corresponding number of released hydrogens and a proportionally higher electronic signal.

DNA Nanoball Sequencing

DNA nanoball sequencing is a type of high throughput sequencing technology used to determine the entire genomic sequence of an organism. The company Complete Genomics uses this technology to sequence samples that researchers submit from several projects. The method uses rolling circle replication to amplify small fragments of genomic DNA into DNA nanoballs. Unchained sequencing by ligation is then used to determine the nucleotide sequence. This method of DNA sequencing allows large numbers of DNA nanoballs to be sequenced per run and at low reagent costs compared to other next generation sequencing platforms. However, only short sequences of

DNA are determined from each DNA nanoball which makes mapping the short reads to a reference genome difficult. This technology has been used for multiple genome sequencing projects and is scheduled to be used for more.

Helioscope(TM) Single Molecule Sequencing

Based on "true single molecule sequencing" technology, Helioscope sequencing uses DNA fragments with added polyA tail adapters, which are attached to the flow cell surface. The next steps involve extension-based sequencing with cyclic washes of the flow cell with fluorescently labeled nucleotides (one nucleotide type at a time, as with the Sanger method). The reads are performed by the Helioscope sequencer. The reads are short, up to 55 bases per run, but recent improvemend of the methodology allowes more accurate reads of homopolymers (stretches of one type of nucleotides) and RNA sequencing.

Single Molecule SMRT(TM) sequencing

SMRT sequencing is based on the sequencing by synthesis approach. The DNA is synthesisd in so calles zero-mode wave-guides (ZMWs) - small well-like containers with the capturing tools located at the bottom of the well. The sequencing is performed with use of unmodified polymerase (attached to the ZMW bottom) and fluorescently labelled nucleotides flowing freely in the solution. The wells are constructed in a way that only the fluorescence occurring by the bottom of the well is detected.

The fluorescent label is detached from the nucleotide at its incorporation into the DNA strand, leaving an unmodified DNA strand. According to Pacific Biosciences, the SMTR technology developer, this methodology allows detection of nucleotide modifications (such ad cytosine methylation). This happens through the observation of polymerase kinetics. This approach allows reads of 1000 nucleotides.

Single Molecule Real Time (RNAP) Sequencing

This method is based on RNA polymerase (RNAP), which is attached to a polystyrene bead, with distal end of sequenced DNA is attached to another bead, with both beads being placed in optical traps. RNAP motion during transcription brings the beads in closer and their relative distance changes, which can then be recorded at a single nucleotide resolution. The sequence is deduced based on the four readouts with lowered concentrations of each of the four nucleotide types (similarly to Sangers method).

Nanopore DNA Sequencing

This method is based on the readout of electrical signal occurring at nucleotides passing by alpha-hemolysin pores covalently bound with cyclodextrin. The DNA passing through the nanopore changes its ion current.

This change is dependent on the shape, size and length of the DNA sequence. Each type of the nucleotide blocks the ion flow through the pore for a different period of time. The method has a potential of development as it does not require modified nucleotides, however single nucleotide resolution is not yet available.

VisiGen Biotechnologies Approach

VisiGen Biotechnologies introduced a specially engineered DNA polymerase for use in their sequencing. This polymerase acts as a sensor - having incorporated a donor fluorescent dye by its active centre. This donor dye acts by FRET (fluorescent resonant energy transfer), inducing fluorescence of differently labeled nucleotides.

This approach allows reads performed at the spead at which polymerase incorporates nucleotides into the sequence (several hundred per second). The nucleotide fluorochrome is released after the incorporation into the DNA strand. The expected read lengths in this approach should reach 1000 nucleotides, however this will have to be confirmed.

Future Methods

Sequencing by hybridization is a non-enzymatic method that uses a DNA microarray. A single pool of DNA whose sequence is to be determined is fluorescently labeled and hybridized to an array containing known sequences. Strong hybridization signals from a given spot on the array identifies its sequence in the DNA being sequenced. Mass spectrometry may be used to determine mass differences between DNA fragments produced in chain-termination reactions.

DNA sequencing methods currently under development include labeling the DNA polymerase, reading the sequence as a DNA strand transits through nanopores, and microscopy-based techniques, such as AFM or transmission electron microscopy that are used to identify the positions of individual nucleotides within long DNA fragments (>5,000 bp) by nucleotide labeling with heavier elements (e.g., halogens) for visual detection and recording.

Third generation technologies aim to increase throughput and decrease the time to result and cost by eliminating the need for excessive reagents and harnessing the processivity of DNA polymerase.

In microfluidic Sanger sequencing the entire thermocycling amplification of DNA fragments as well as their separation by electrophoresis is done on a single glass wafer (approximately 10 cm in diameter) thus reducing the reagent usage as well as cost.

In some instances researchers have shown that they can increase the throughput of conventional sequencing through the use of microchips. Research will still need to be done in order to make this use of technology effective.

In October 2006, the X Prize Foundation established an initiative to promote the development of full genome sequencing technologies, called the Archon X Prize, intending to award $10 million to "the first Team that can build a device and use it to sequence 100 human genomes within 10 days or less, with an accuracy of no more than one error in every 100,000 bases sequenced, with sequences accurately covering at least 98% of the genome, and at a recurring cost of no more than $10,000 (US) per genome."

Each year NHGRI promotes grants for new research and developments in genomics. 2010 grants and 2011 candidates include continuing work in microfluidic, polony and base-heavy sequencing methodologies

Major Landmarks in DNA Sequencing

- 1953 Discovery of the structure of the DNA double helix.
- 1972 Development of recombinant DNA technology, which permits isolation of defined fragments of DNA; prior to this, the only accessible samples for sequencing were from bacteriophage or virus DNA.
- 1977 The first complete DNA genome to be sequenced is that of bacteriophage öX174.
- 1977 Allan Maxam and Walter Gilbert publish "DNA sequencing by chemical degradation". Frederick Sanger, independently,

publishes "DNA sequencing with chain-terminating inhibitors".

- 1984 Medical Research Council scientists decipher the complete DNA sequence of the Epstein-Barr virus, 170 kb.
- 1986 Leroy E. Hood's laboratory at the California Institute of Technology and Smith announce the first semi-automated DNA sequencing machine.
- 1987 Applied Biosystems markets first automated sequencing machine, the model ABI 370.
- 1990 .The U.S. National Institutes of Health (NIH) begins large-scale sequencing trials on Mycoplasma capricolum, Escherichia coli, Caenorhabditis elegans, and Saccharomyces cerevisiae (at US$0.75/base).
- 1991 Sequencing of human expressed sequence tags begins in Craig Venter's lab, an attempt to capture the coding fraction of the human genome.
- 1995 Craig Venter, Hamilton Smith, and colleagues at The Institute for Genomic Research (TIGR) publish the first complete genome of a free-living organism, the bacterium Haemophilus influenzae. The circular chromosome contains 1,830,137 bases and its publication in the journal Science marks the first use of whole-genome shotgun sequencing, eliminating the need for initial mapping efforts.
- 1996 Pål Nyrén and his student Mostafa Ronaghi at the Royal Institute of Technology in Stockholm publish their method of pyrosequencing
- 1998 Phil Green and Brent Ewing of the University of Washington publish "phred" for sequencer data analysis.
- 2000 Lynx Therapeutics publishes and markets "MPSS" - a parallelized, adapter/ligation-mediated, bead-based sequencing technology, launching "next-generation" sequencing.
- 2001 A draft sequence of the human genome is published.
- 2004 454 Life Sciences markets a parallelized version of pyrosequencing. The first version of their machine reduced sequencing costs 6-fold compared to automated Sanger sequencing, and was the second of a new generation of sequencing technologies, after MPSS.

Genome

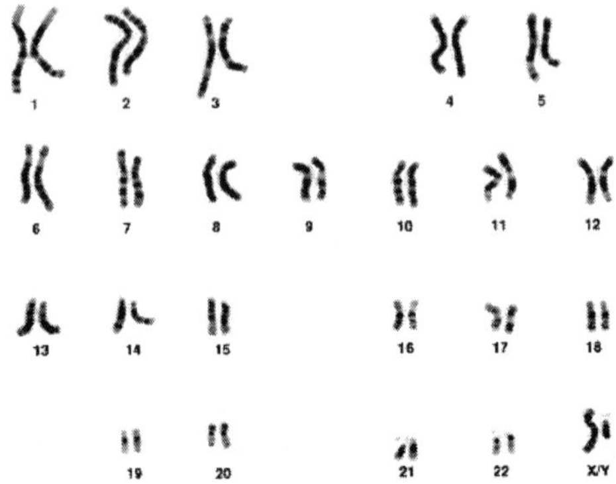

Fig. 9: An image of the 46 chromosomes, making up the diploid genome of human male. (The mitochondrial chromosome is not shown.)

In modern molecular biology and genetics, the genome is the entirety of an organism's hereditary information. It is encoded either in DNA or, for many types of virus, in RNA. The genome includes both the genes and the non-coding sequences of the DNA/RNA.

Origin of Term

The term was adapted in 1920 by Hans Winkler, Professor of Botany at the University of Hamburg, Germany. In Greek, the word genome (ãßíìáé) means "I become, I am born, to come into being". The Oxford English Dictionary suggests the name to be a blend of the words gene and chromosome. A few related -ome words already existed, such asbiome and rhizome, forming a vocabulary into which genome fits systematically.

Overview

Some organisms have multiple copies of chromosomes, diploid, triploid, tetraploid and so on. In classical genetics, in a sexually reproducing organism (typically eukarya) thegamete has half the number of chromosomes of the somatic cell and the genome is a full set of chromosomes in a gamete. In haploid organisms, including cells of bacteria,archaea, and in organelles including mitochondria and chloroplasts, or viruses, that similarly contain genes, the single or set of circular and/or linear chains of DNA (or RNA forsome viruses), likewise constitute the genome. The term genome can be applied specifically to mean that stored on a complete set of nuclear DNA (i.e., the "nuclear genome") but can also be applied to that stored within organelles that contain their own DNA, as with the "mitochondrial genome" or the "chloroplast genome". Additionally, the genome can comprise nonchromosomal genetic elements such as viruses, plasmids, and transposable elements.

When people say that the genome of a sexually reproducing species has been "sequenced", typically they are referring to a determination of the sequences of one set ofautosomes and one of each type of sex chromosome, which together represent both of the possible sexes. Even in species that exist in only one sex, what is described as "a genome sequence" may be a composite read from the chromosomes of various individuals. In general use, the phrase "genetic makeup" is sometimes used conversationally to mean the genome of a particular individual or organism. The study of the global properties of genomes of related organisms is usually referred to asgenomics, which distinguishes it from genetics which generally studies the properties of single genes or groups of genes.

Both the number of base pairs and the number of genes vary widely from one species to another, and there is only a rough correlation between the two (an observation known as the C-value paradox). At present, the highest known number of genes is around 60,000, for the protozoan causingtrichomoniasis.

Types

Most biological entities that are more complex than a virus sometimes or always carry additional genetic material besides that which resides in their chromosomes. In some contexts, such as sequencing the genome of a pathogenic microbe, "genome" is meant to include information stored on this auxiliary material, which is carried in plasmids. In such circumstances then, "genome" describes all of the genes and information on non-coding DNA that have the potential to be present.

In eukaryotes such as plants, protozoa and animals, however, "genome" carries the typical connotation of only information on chromosomal DNA. So although these organisms contain chloroplastsand/or mitochondria that have their own DNA, the genetic information contained by DNA within these organelles is not considered part of the genome. In fact, mitochondria are sometimes said to have their own genome often referred to as the "mitochondrial genome". The DNA found within the chloroplast may be referred to as the "plastome".

Chapter 4

Genomes and Genetic Variation

A genome does not capture the genetic diversity or the genetic polymorphism of a species. For example, the human genome sequence in principle could be determined from just half the information on the DNA of one cell from one individual. To learn what variations in genetic information underlie particular traits or diseases requires comparisons across individuals.

This point explains the common usage of "genome" (which parallels a common usage of "gene") to refer not to the information in any particular DNA sequence, but to a whole family of sequences that share a biological context.

Although this concept may seem counter intuitive, it is the same concept that says there is no particular shape that is the shape of a cheetah. Cheetahs vary, and so do the sequences of their genomes. Yet both the individual animals and their sequences share commonalities, so one can learn something about cheetahs and "cheetah-ness" from a single example of either.

Sequencing and Mapping

The Human Genome Project was organized to map and to sequence the human genome. Other genome projects include mouse, rice, the plant Arabidopsis thaliana, the puffer fish, and bacteria like E. coli. In 1976, Walter Fiers at the University of Ghent (Belgium) was the first to establish the complete nucleotide sequence of a viral RNA-genome (bacteriophage MS2).

The first DNA-genome project to be completed was the Phage Ö-X174, with only 5386 base pairs, which was sequenced by Fred Sanger in 1977. The first bacterial genome to be completed was that of Haemophilus influenzae, completed by a team at The Institute for

Genomic Research in 1995. A few months later, the first eukaryotic genome was completed, with the 16 chromosomes of budding yeast Saccharomyces cerevisiae being released as a result of a European-led effort begun in the mid-1980s.

The development of new technologies has made it dramatically easier and cheaper to do sequencing, and the number of complete genome sequences is growing rapidly. Among many genome databases, the one maintained by the US National Institutes of Health is inclusive.

These new technologies open up the prospect of personal genome sequencing as an important diagnostic tool. A major step toward that goal was the completion of the decipherment of the full genomeof DNA pioneer James D. Watson in 2007.

Whereas a genome sequence lists the order of every DNA base in a genome, a genome map identifies the landmarks. A genome map is less detailed than a genome sequence and aids in navigating around the genome. A fundamental step in the Human genome project was the release of a detailed genomic map by Jean Weissenbach and his team at the Genoscope in Paris .

Genome Evolution

Genomes are more than the sum of an organism's genes and have traits that may be measured and studied without reference to the details of any particular genes and their products. Researchers compare traits such as chromosome number (karyotype), genome size, gene order, codon usage bias, and GC-content to determine what mechanisms could have produced the great variety of genomes that exist today.

Duplications play a major role in shaping the genome. Duplications may range from extension of short tandem repeats, to duplication of a cluster of genes, and all the way to duplications of entire chromosomes or even entire genomes. Such duplications are probably fundamental to the creation of genetic novelty.

Horizontal gene transfer is invoked to explain how there is often extreme similarity between small portions of the genomes of two organisms that are otherwise very distantly related. Horizontal gene transfer seems to be common among many microbes. Also, eukaryotic cells seem to have experienced a transfer of some genetic material from their chloroplast and mitochondrial genomes to their nuclear chromosomes.

Heredity

Heredity is the passing of traits to offspring (from its parent or ancestors). This is the process by which an offspring cell or organism acquires or becomes predisposed to the characteristics of its parent cell or organism. Through heredity, variations exhibited by individuals can accumulate and cause some species to evolve. The study of heredity inbiology is called genetics, which includes the field of epigenetics.

Overview

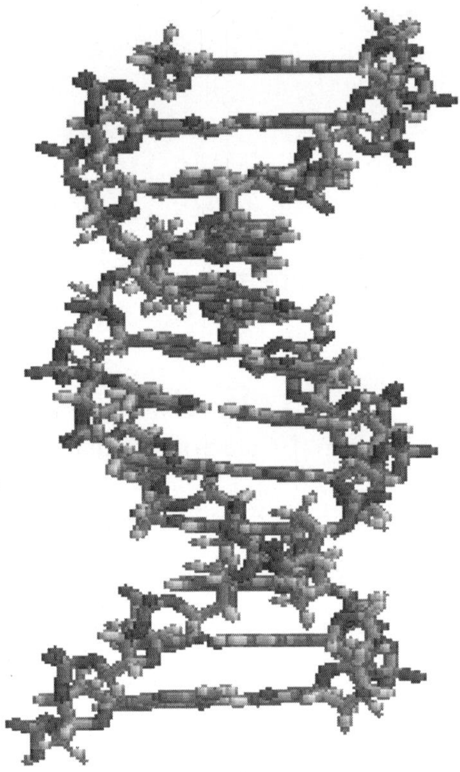

FIg. 1: DNA structure. Bases are in the centre, surrounded by phosphate–sugar chains in adouble helix.

In humans, eye color is an inherited characteristic and an individual might inherit the "brown-eye trait" from one of the parents. Inherited traits are controlled by genes and the complete set of genes within an organism's genome is called its genotype.

The complete set of observable traits that make up the structure and behaviour of an organism is called its phenotype. These traits come from the interaction of its genotype with the environment. As a result, many aspects of an organism's phenotype are not inherited. For example,suntanned skin comes from the interaction between a person's genotype and sunlight; thus, suntans are not passed on to people's children. However, some people tan more easily than others, due to differences in their genotype; a striking example are people with the inherited trait ofalbinism, who do not tan at all and are very sensitive to sunburn.

Heritable traits are known to be passed from one generation to the next via DNA, a molecule that encodes genetic information. DNA is a longpolymer composed of four types of bases. The sequence of bases along a particular DNA molecule specify the genetic information, in a manner similar to a sequence of letters spelling out a word. Before a cell divides, the DNA is copied, so that each of the resulting two cells will inherit the DNA sequence.

Portions of a DNA molecule that specify a single functional unit are called genes; different genes have different sequences of bases. Within cells, the long strands of DNA form condensed structures called chromosomes. The specific location of a DNA sequence within a chromosome is known as a locus. If the DNA sequence at a locus varies between individuals, the different forms of this sequence are called alleles. DNA sequences can change through mutations, producing new alleles. If a mutation occurs within a gene, the new allele may affect the trait that the gene controls, altering the phenotype of the organism.

However, while this simple correspondence between an allele and a trait works in some cases, most traits are more complex and are controlled by multiple interacting genes within and among organisms. Developmental biologists suggest that complex interactions in genetic networks and communication among cells can lead to heritable variations that may underlay some of the mechanics in developmental plasticity and canalization.

Recent findings have confirmed important examples of heritable changes that cannot be explained by direct agency of the DNA molecule. These phenomena are classed as epigenetic inheritance systems that are causally or independently evolving over genes. Research into modes and mechanisms of epigenetic inheritance is still in its scientific infancy, however, this area of research has attracted much recent activity as it broadens the scope of heritability and evolutionary biology in general.

DNA methylation marking chromatin, self-sustaining metabolic loops, gene silencing by RNA interference, and the three dimensional conformation of proteins (such as prions) are areas where epigenetic inheritance systems have been discovered at the organismic level.

Heritability may also occur at even larger scales. For example, ecological inheritance through the process of niche construction is defined by the regular and repeated activities of organisms in their environment. This generates a legacy of effect that modifies and feeds back into the selection regime of subsequent generations. Descendants inherit genes plus environmental characteristics generated by the ecological actions of ancestors.

Other examples of heritability in evolution that are not under the direct control of genes include the inheritance of cultural traits, group heritability, and symbiogenesis. These examples of heritability that operate above the gene are covered broadly under the title of multilevel or hierarchical selection, which has been a subject of intense debate in the history of evolutionary science.

Relation to Theory of Evolution

When Charles Darwin proposed his theory of evolution in 1859, one of its major problems was the lack of an underlying mechanism for heredity. Darwin believed in a mix of blending inheritance and the inheritance of acquired traits (pangenesis). Blending inheritance would lead to uniformity across populations in only a few generations and thus would remove variation from a population on which natural selection could act.

This led to Darwin adopting some Lamarckian ideas in later editions of On the Origin of Species and his later biological works. Darwin's primary approach to heredity was to outline how it appeared to work (noticing that traits could be inherited which were not expressed explicitly in the parent at the time of reproduction, that certain traits could be sex-linked, etc.) rather than suggesting mechanisms.

Darwin's initial model of heredity was adopted by, and then heavily modified by, his cousin Francis Galton, who laid the framework for the biometric school of heredity. Galton rejected the aspects of Darwin's pangenesis model which relied on acquired traits.

The inheritance of acquired traits was shown to have little basis in the 1880s when August Weismann cut the tails off many generations of mice and found that their offspring continued to develop tails.

History

The ancients had a variety of ideas about heredity: Theophrastus proposed that male flowers caused female flowers to ripen; Hippocrates speculated that "seeds" were produced by various body parts and transmitted to offspring at the time of conception, and Aristotle thought that male and female semen mixed at conception. Aeschylus, in 458 BC, proposed the male as the parent, with the female as a "nurse for the young life sown within her." Various hereditary mechanisms were envisaged without being properly tested or quantified. These included blending inheritance and the inheritance of acquired traits. Nevertheless, people were able to develop domestic breeds of animals as well as crops through artificial selection. The inheritance of acquired traits also formed a part of early Lamarckian ideas on evolution.

During the 18th century, Dutch microscopist Antonie van Leeuwenhoek (1632–1723) discovered "animalcules" in the sperm of humans and other animals. Some scientists speculated they saw a "little man" (homunculus) inside each sperm. These scientists formed a school of thought known as the "spermists." They contended the only contributions of the female to the next generation were the womb in which the homunculus grew, and prenatal influences of the womb. An opposing school of thought, the ovists, believed that the future human was in the egg, and that sperm merely stimulated the growth of the egg. Ovists thought women carried eggs containing boy and girl children, and that the gender of the offspring was determined well before conception.

Gregor Mendel: Father of Genetics

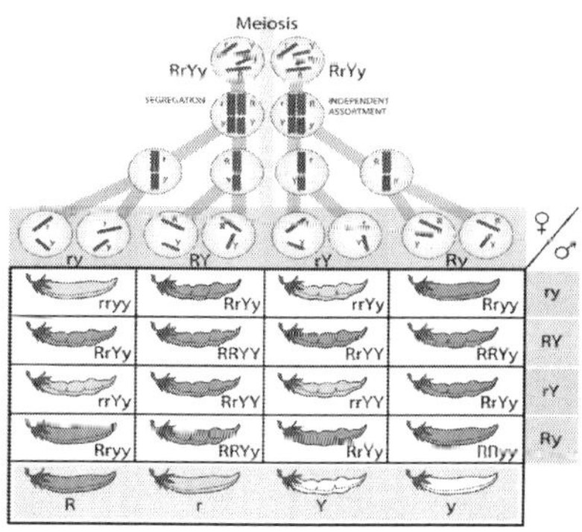

Fig. 2: Table showing how the genes exchange according to segregation or independent assortment during meiosis and how this translates into Mendel's laws

The idea of particulate inheritance of genes can be attributed to the Moravian monk Gregor Mendel who published his work on pea plants in 1865. However, his work was not widely known and was rediscovered in 1901. It was initially assumed the Mendelian inheritance only accounted for large (qualitative) differences, such as those seen by Mendel in his pea plants—and the idea of additive effect of (quantitative) genes was not realised until R.A. Fisher's (1918) paper, "The Correlation Between Relatives on the Supposition of Mendelian Inheritance" Mendel's overall contribution gave scientists an useful overview that traits were inheritable. As of today, his pea plant demonstration became the foundation of the study of Mendelian Traits. These traits can be traced on a single loci.

Modern Development of Genetics and Heredity

In the 1930s, work by Fisher and others resulted in a combination of Mendelian and biometric schools into the modern evolutionary synthesis. The modern synthesis bridged the gap between experimental geneticists and naturalists; and between both and palaeontologists, stating that:

1. All evolutionary phenomena can be explained in a way consistent with known genetic mechanisms and the observational evidence of naturalists.
2. Evolution is gradual: small genetic changes, recombination ordered by natural selection. Discontinuities amongst species (or other taxa) are explained as originating gradually through geographical separation and extinction (not saltation).
3. Selection is overwhelmingly the main mechanism of change; even slight advantages are important when continued. The object of selection is the phenotype in its surrounding environment. The role of genetic drift is equivocal; though strongly supported initially by Dobzhansky, it was downgraded later as results from ecological genetics were obtained.
4. The primacy of population thinking: the genetic diversity carried in natural populations is a key factor in evolution. The strength of natural selection in the wild was greater than expected; the effect of ecological factors such as niche occupation and the significance of barriers to gene flow are all important.
5. In palaeontology, the ability to explain historical observations by extrapolation from micro to macro-evolution is proposed.

Historical contingency means explanations at different levels may exist. Gradualism does not mean constant rate of change.

The idea that speciation occurs after populations are reproductively isolated has been much debated. In plants, polyploidy must be included in any view of speciation. Formulations such as 'evolution consists primarily of changes in the frequencies of alleles between one generation and another' were proposed rather later.

The traditional view is that developmental biology ('evo-devo') played little part in the synthesis, but an account of Gavin de Beer's work by Stephen Jay Gould suggests he may be an exception.

Almost all aspects of the synthesis have been challenged at times, with varying degrees of success. There is no doubt, however, that the synthesis was a great landmark in evolutionary biology. It cleared up many confusions, and was directly responsible for stimulating a great deal of research in the post-World War II era.

Trofim Lysenko however caused a backlash of what is now called Lysenkoism in the Soviet Union when he emphasised Lamarckian ideas on the inheritance of acquired traits. This movement affected agricultural research and led to food shortages in the 1960s and seriously affected the USSR.

Common Genetic Disorders

- Down syndrome
- Huntington's disease
- Phenylketonuria (PKU)
- FOXP2
- Hemophilia

Types of Heredity

Dominant and Recessive

An allele is said to be dominant if it is always expressed in the appearance of an organism (phenotype). For example, in peas the allele for green pods, G, is dominant to that for yellow pods, g. Since the allele for green pods is dominant, pea plants with the pair of alleles GG (homozygote) or Gg (heterozygote) will have green pods. The allele for yellow pods is recessive. The effects of this allele are only seen when it is present in both chromosomes, gg (homozygote).

The description of a mode of biological inheritance consists of three main categories:

1. Number of Involved loci

- Monogenetic (also called "simple") – one locus
- Oligogenetic – few loci
- Polygenetic – many loci

2. Involved Chromosomes

- Autosomal – loci are not situated on a sex chromosome
- Gonosomal – loci are situated on a sex chromosome
- X-chromosomal – loci are situated on the X-chromosome (the more common case)
- Y-chromosomal – loci are situated on the Y-chromosome
- Mitochondrial – loci are situated on the mitochondrial DNA

3. Correlation Genotype–phenotype

- Dominant
- Intermediate (also called "codominant")
- Recessive

These three categories are part of every exact description of a mode of inheritance in the above order. Additionally, more specifications may be added as follows:

4. Coincidental and Environmental Interactions

- Penetrance
- Complete
- Incomplete (percentual number)
- Expressivity
- Invariable
- Variable
- Heritability (in polygenetic and sometimes also in oligogenetic modes of inheritance)
- Maternal or paternal imprinting phenomena.

5. Sex-linked Interactions

- Sex-linked inheritance (gonosomal loci)
- Sex-limited phenotype expression (e.g., cryptorchism)

- Inheritance through the maternal line (in case of mitochondrial DNA loci)
- Inheritance through the paternal line (in case of Y-chromosomal loci)

6. Locus–locus Interactions

- Epistasis with other loci (e.g., overdominance)
- Gene coupling with other loci
- Homozygotous lethal factors
- Semi-lethal factors

Determination and description of a mode of inheritance is primarily achieved through statistical analysis of pedigree data. In case the involved loci are known, methods of molecular genetics can also be employed.

Mutation

In molecular biology and genetics, mutations are changes in a genomic sequence: the DNA sequence of a cell's genome or the DNA or RNAsequence of a virus. They can be defined as sudden and spontaneous changes in the cell. Mutations are caused by radiation, viruses,transposons and mutagenic chemicals, as well as errors that occur during meiosis or DNA replication. They can also be induced by the organism itself, by cellular processes such as hypermutation. A mutation is passed to the offspring stably, unless it is a dynamic mutation.

Mutation can result in several different types of change in sequences;(DNA) these can either have no effect, alter the product of a gene, or prevent the gene from functioning properly or completely. Studies in the fly Drosophila melanogaster suggest that if a mutation changes a proteinproduced by a gene, this will probably be harmful, with about 70 percent of these mutations having damaging effects, and the remainder being either neutral or weakly beneficial. Due to the damaging effects that mutations can have on genes, organisms have mechanisms such as DNA repair to remove mutations.

Viruses that use RNA as their genetic material have rapid mutation rates, which can be an advantage since these viruses will evolve constantly and rapidly, and thus evade the defensive responses of e.g. the human immune system.

Description

Mutations can involve large sections of DNA becoming duplicated, usually through genetic recombination. These duplications are a major source of raw material for evolving new genes, with tens to hundreds of genes duplicated in animal genomes every million years. Most genes belong to larger families of genes of shared ancestry. Novel genes are produced by several methods, commonly through the duplication and mutation of an ancestral gene, or by recombining parts of different genes to form new combinations with new functions.

Here, domains act as modules, each with a particular and independent function, that can be mixed together to produce genes encoding new proteins with novel properties. For example, the human eye uses four genes to make structures that sense light: three for color vision and one for night vision; all four arose from a single ancestral gene. Another advantage of duplicating a gene (or even anentire genome) is that this increases redundancy; this allows one gene in the pair to acquire a new function while the other copy performs the original function. Other types of mutation occasionally create new genes from previously noncoding DNA.

Changes in chromosome number may involve even larger mutations, where segments of the DNA within chromosomes break and then rearrange. For example, in the Homininae, two chromosomes fused to produce human chromosome 2; this fusion did not occur in the lineage of the other apes, and they retain these separate chromosomes. In evolution, the most important role of such chromosomal rearrangements may be to accelerate the divergence of a population into new species by making populations less likely to interbreed, and thereby preserving genetic differences between these populations.

Sequences of DNA that can move about the genome, such as transposons, make up a major fraction of the genetic material of plants and animals, and may have been important in the evolution of genomes. For example, more than a million copies of the Alu sequence are present in the human genome, and these sequences have now been recruited to perform functions such as regulatinggene expression. Another effect of these mobile DNA sequences is that when they move within a genome, they can mutate or delete existing genes and thereby produce genetic diversity.

Nonlethal mutations accumulate within the gene pool and increase the amount of genetic variation. The abundance of some genetic changes within the gene pool can be reduced by natural selection, while other "more favorable" mutations may accumulate and result in adaptive changes.

For example, a butterfly may produce offspring with new mutations. The majority of these mutations will have no effect; but one might change the color of one of the butterfly's offspring, making it harder (or easier) for predators to see. If this color change is advantageous, the chance of this butterfly surviving and producing its own offspring are a little better, and over time the number of butterflies with this mutation may form a larger percentage of the population.

Neutral mutations are defined as mutations whose effects do not influence the fitness of an individual. These can accumulate over time due to genetic drift. It is believed that the overwhelming majority of mutations have no significant effect on an organism's fitness. Also, DNA repair mechanisms are able to mend most changes before they become permanent mutations, and many organisms have mechanisms for eliminating otherwise permanently mutated somatic cells.

Mutation is generally accepted by biologists as the mechanism by which natural selection acts, generating advantageous new traits that survive and multiply in offspring as well as disadvantageous traits, in less fit offspring, that tend to die out.

Causes

Two classes of mutations are spontaneous mutations (molecular decay) and induced mutations caused by mutagens.

Spontaneous Mutation

Spontaneous mutations on the molecular level can be caused by:

- Tautomerism – A base is changed by the repositioning of a hydrogen atom, altering the hydrogen bonding pattern of that base resulting in incorrect base pairing during replication.
- Depurination – Loss of a purine base (A or G) to form an apurinic site (AP site).
- Deamination – Hydrolysis changes a normal base to an atypical base containing a keto group in place of the original amine group. Examples include C '! U and A '! HX (hypoxanthine),

which can be corrected by DNA repair mechanisms; and 5MeC (5-methylcytosine) '! T, which is less likely to be detected as a mutation because thymine is a normal DNA base.

- Slipped strand mispairing - Denaturation of the new strand from the template during replication, followed by renaturation in a different spot ("slipping"). This can lead to insertions or deletions.

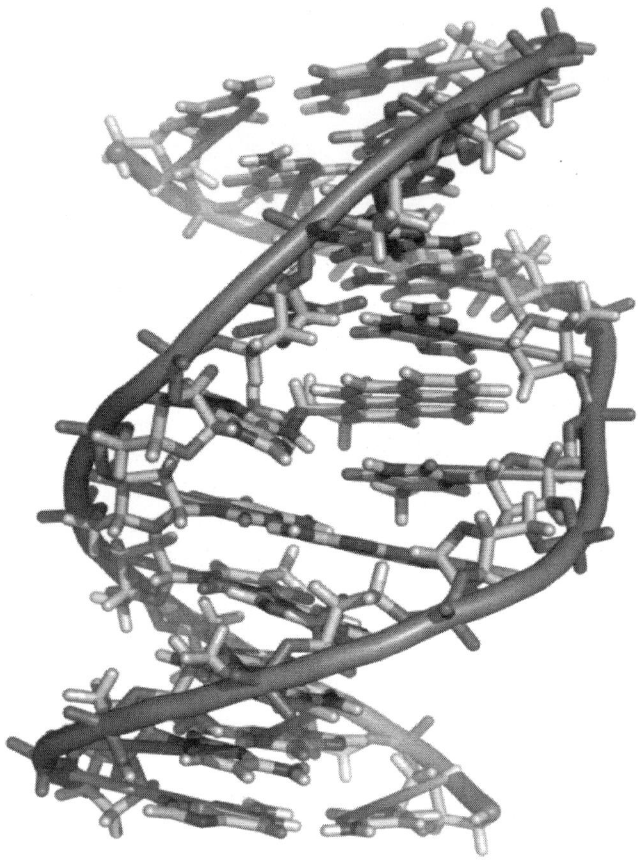

Fig. 3: A covalent adduct between benzo[a]pyrene,the major mutagen in tobacco smoke, and DNA

Induced Mutation

Induced mutations on the molecular level can be caused by:

- Chemicals
  - Hydroxylamine NH_2OH

  - Base analogs (e.g. BrdU)
  - Alkylating agents (e.g. N-ethyl-N-nitrosourea) These agents can mutate both replicating and non-replicating DNA. In contrast, a base analog can only mutate the DNA when the analog is incorporated in replicating the DNA. Each of these classes of chemical mutagens has certain effects that then lead to transitions, transversions, or deletions.
  - Agents that form DNA adducts (e.g. ochratoxin A metabolites)
  - DNA intercalating agents (e.g. ethidium bromide)
  - DNA crosslinkers
  - Oxidative damage
  - Nitrous acid converts amine groups on A and C to diazo groups, altering their hydrogen bonding patterns which leads to incorrect base pairing during replication.
- Radiation
  - Ultraviolet radiation (nonionizing radiation). Two nucleotide bases in DNA – cytosine and thymine – are most vulnerable to radiation that can change their properties. UV light can induce adjacent pyrimidine bases in a DNA strand to become covalently joined as a pyrimidine dimer. UV radiation, particularly longer-wave UVA, can also cause oxidative damage to DNA.
  - Ionizing radiation
  - Radioactive decay, such as C in DNA
- Viral infections

DNA has so-called hotspots, where mutations occur up to 100 times more frequently than the normal mutation rate. A hotspot can be at an unusual base, e.g., 5-methylcytosine.

Mutation rates also vary across species. Evolutionary biologists have theorized that higher mutation rates are beneficial in some situations, because they allow organisms to evolve and therefore adapt more quickly to their environments. For example, repeated exposure of bacteria to antibiotics, and selection of resistant mutants, can result in the selection of bacteria that have a much higher mutation rate than the original population (mutator strains).

Classification of Mutation Types

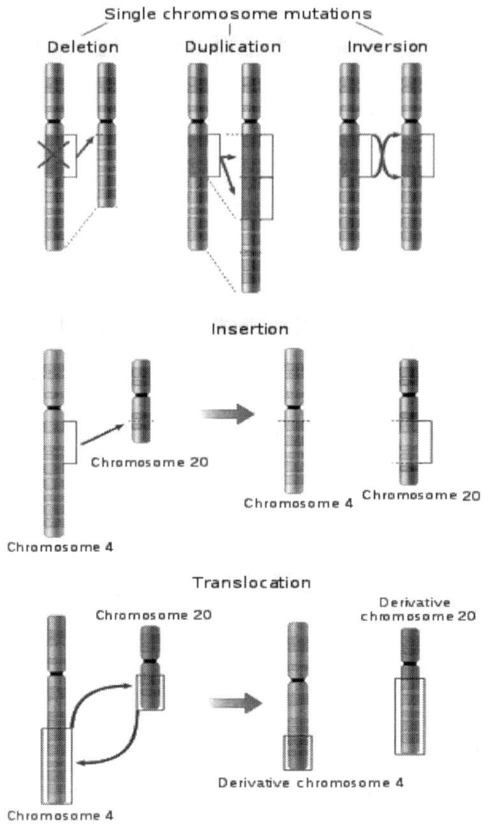

Fig. 4: Illustrations of five types of chromosomal mutations.

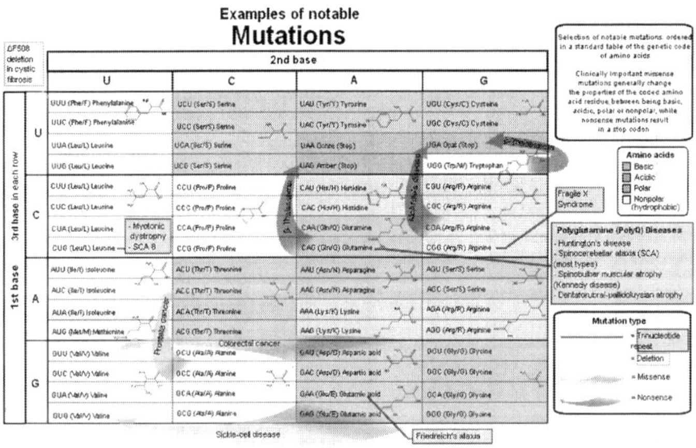

Fig. 5: Selection of disease-causing mutations, in a standard table of the genetic code of amino acids.

By Effect on Structure

The sequence of a gene can be altered in a number of ways. Gene mutations have varying effects on health depending on where they occur and whether they alter the function of essential proteins. Mutations in the structure of genes can be classified as:

- Small-scale mutations, such as those affecting a small gene in one or a few nucleotides, including:
- Point mutations, often caused by chemicals or malfunction of DNA replication, exchange a single nucleotide for another. These changes are classified as transitions or transversions. Most common is the transition that exchanges a purine for a purine (A "! G) or a pyrimidine for a pyrimidine, (C "! T). A transition can be caused by nitrous acid, base mis-pairing, or mutagenic base analogs such as5-bromo-2-deoxyuridine (BrdU). Less common is a transversion, which exchanges a purine for a pyrimidine or a pyrimidine for a purine (C/T "! A/G). An example of a transversion is adenine (A) being converted into a cytosine (C). A point mutation can be reversed by another point mutation, in which the nucleotide is changed back to its original state (true reversion) or by second-site reversion (a complementary mutation elsewhere that results in regained gene functionality). Point mutations that occur within the protein coding region of a gene may be classified into three kinds, depending upon what the erroneous codon codes for:
- Silent mutations: which code for the same amino acid.
- Missense mutations: which code for a different amino acid.
- Nonsense mutations: which code for a stop and can truncate the protein.
- Insertions add one or more extra nucleotides into the DNA. They are usually caused by transposable elements, or errors during replication of repeating elements (e.g. AT repeats). Insertions in the coding region of a gene may alter splicing of the mRNA(splice site mutation), or cause a shift in the reading frame (frameshift), both of which can significantly alter the gene product. Insertions can be reverted by excision of the transposable element.
- Deletions remove one or more nucleotides from the DNA. Like insertions, these mutations can alter the reading frame of the

gene. They are generally irreversible: though exactly the same sequence might theoretically be restored by an insertion, transposable elements able to revert a very short deletion (say 1–2 bases) in any location are either highly unlikely to exist or do not exist at all. Note that a deletion is not the exact opposite of an insertion: the former is quite random while the latter consists of a specific sequence inserting at locations that are not entirely random or even quite narrowly defined.

- Large-scale mutations in chromosomal structure, including:
- Amplifications (or gene duplications) leading to multiple copies of all chromosomal regions, increasing the dosage of the genes located within them.
- Deletions of large chromosomal regions, leading to loss of the genes within those regions.
- Mutations whose effect is to juxtapose previously separate pieces of DNA, potentially bringing together separate genes to form functionally distinct fusion genes (e.g. bcr-abl). These include:
- Chromosomal translocations: interchange of genetic parts from nonhomologous chromosomes.
- Interstitial deletions: an intra-chromosomal deletion that removes a segment of DNA from a single chromosome, thereby apposing previously distant genes. For example, cells isolated from a human astrocytoma, a type of brain tumor, were found to have a chromosomal deletion removing sequences between the "fused in glioblastoma"gene and the receptor tyrosine kinase "ros", producing a fusion protein (FIG-ROS). The abnormal FIG-ROS fusion protein has constitutively active kinase activity that causes oncogenic transformation (a transformation from normal cells to cancer cells).
- Chromosomal inversions: reversing the orientation of a chromosomal segment.
- Loss of heterozygosity: loss of one allele, either by a deletion or recombination event, in an organism that previously had two different alleles.

By Effect on Function

- Loss-of-function mutations are the result of gene product having less or no function. When the allele has a complete loss of function (null allele) it is often called an amorphic mutation. Phenotypes associated with such mutations are most often recessive. Exceptions are when the organism is haploid, or when the reduced dosage of a normal gene product is not enough for a normal phenotype (this is calledhaploinsufficiency).

- Gain-of-function mutations change the gene product such that it gains a new and abnormal function. These mutations usually have dominant phenotypes. Often called a neomorphic mutation.
- Dominant negative mutations (also called antimorphic mutations) have an altered gene product that acts antagonistically to the wild-type allele. These mutations usually result in an altered molecular function (often inactive) and are characterised by a dominant or semi-dominant phenotype. In humans, Marfan syndrome is an example of a dominant negative mutation occurring in anautosomal dominant disease. In this condition, the defective glycoprotein product of the fibrillin gene (FBN1) antagonizes the product of the normal allele.
- Lethal mutations are mutations that lead to the death of the organisms which carry the mutations.
- A back mutation or reversion is a point mutation that restores the original sequence and hence the original phenotype.

By Effect on Fitness

In applied genetics it is usual to speak of mutations as either harmful or beneficial.

- A harmful mutation is a mutation that decreases the fitness of the organism.
- A beneficial mutation is a mutation that increases fitness of the organism, or which promotes traits that are desirable.

In theoretical population genetics, it is more usual to speak of such mutations as deleterious or advantageous. In the neutral theory of molecular evolution, genetic drift is the basis for most variation at the molecular level.

- A neutral mutation has no harmful or beneficial effect on the organism. Such mutations occur at a steady rate, forming the basis for the molecular clock.
- A deleterious mutation has a negative effect on the phenotype, and thus decreases the fitness of the organism.

- An advantageous mutation has a positive effect on the phenotype, and thus increases the fitness of the organism.
- A nearly neutral mutation is a mutation that may be slightly deleterious or advantageous, although most nearly neutral mutations are slightly deleterious.

In reality, viewing the fitness effects of mutations in these discrete categories is an oversimplification. Attempts have been made to infer the distribution of fitness effects using mutagenesis experiments or theoretical models applied to molecular sequence data. However, the current distribution is still uncertain, and some aspects of the distribution likely vary between species.

By Inheritance

- inheritable generic in pro-generic tissue or cells on path to be changed to gametes.
- non inheritable somatic (e.g., carcinogenic mutation)
- non inheritable post mortem aDNA mutation in decaying remains.

By pattern of inheritance The human genome contains two copies of each gene – a paternal and a maternal allele.

- A heterozygous mutation is a mutation of only one allele.
- A homozygous mutation is an identical mutation of both the paternal and maternal alleles.
- Compound heterozygous mutations or a genetic compound comprises two different mutations in the paternal and maternal alleles.
- A wildtype or homozygous non-mutated organism is one in which neither allele is mutated. (Just not a mutation)

By impact on protein sequence

- A frameshift mutation is a mutation caused by insertion or deletion of a number of nucleotides that is not evenly divisible by three from a DNA sequence. Due to the triplet nature of gene expression by codons, the insertion or deletion can disrupt the reading frame, or the grouping of the codons, resulting in a completely different translation from the original. The earlier in the sequence the deletion or insertion occurs, the more altered the protein produced is.

In contrast, any insertion or deletion that is evenly divisible by three is termed an in-frame mutation

- A nonsense mutation is a point mutation in a sequence of DNA that results in a premature stop codon, or a nonsense codon in the transcribed mRNA, and possibly a truncated, and often nonfunctional protein product.
- Missense mutations or nonsynonymous mutations are types of point mutations where a single nucleotide is changed to cause substitution of a different amino acid. This in turn can render the resulting protein nonfunctional. Such mutations are responsible for diseases such as Epidermolysis bullosa, sickle-cell disease, and SOD1 mediated ALS (Boillée 2006, p. 39).
- A neutral mutation is a mutation that occurs in an amino acid codon which results in the use of a different, but chemically similar, amino acid. The similarity between the two is enough that little or no change is often rendered in the protein. For example, a change from AAA to AGA will encode arginine, a chemically similar molecule to the intended lysine. Neutral mutations occur because of the degenerate nature of the genetic code.
- Silent mutations are mutations that do not result in a change to the amino acid sequence of a protein. They may occur in a region that does not code for a protein, or they may occur within a codon in a manner that does not alter the final amino acid sequence. The phrase silent mutation is often used interchangeably with the phrase synonymous mutation; however, synonymous mutations are a subcategory of the former, occurring only within exons. The name silent could be a misnomer. For example, a silent mutation in the exon/intron border may lead to alternative splicing by changing the splice site.

By Inheritance Ability

A mutation has caused this garden moss rose to produce flowers of different colors. This is a somatic mutation that may also be passed on in the germ line.

In multicellular organisms with dedicated reproductive cells, mutations can be subdivided into germ line mutations, which can be passed on to descendants through their reproductive cells, and somatic mutations (also called acquired mutations), which involve cells outside the dedicated reproductive group and which are not usually transmitted to descendants.

Fig. 6: A germline mutation gives rise to a constitutional mutation in the offspring, that is, a mutation that is present in every cell. A constitutional mutation can also occur very soon after fertilisation, or continue from a previous constitutional mutation in a parent.

If the organism can reproduce asexually through mechanisms such as cuttings or budding the distinction can become blurred. For example, plants can sometimes transmit somatic mutations to their descendants asexually or sexually where flower buds develop in somatically mutated parts of plants. A new mutation that was not inherited from either parent is called a de novo mutation. The source of the mutation is unrelated to the consequence, although the consequences are related to which cells were mutated.

Special Classes

- Conditional mutation is a mutation that has wild-type (or less severe) phenotype under certain "permissive" environmental conditions and a mutant phenotype under certain "restrictive" conditions. For example, a temperature-sensitive mutation can cause cell death at high temperature (restrictive condition), but might have no deleterious consequences at a lower temperature (permissive condition).

Nomenclature

A committee of the Human Genome Variation Society (HGVS) has developed the standard human sequence variant nomenclature, which should be used by researchers and DNA diagnostic centers to generate unambiguous mutation descriptions. In principle, this nomenclature can also be used to describe mutations in other organisms. The nomenclature specifies the type of mutation and base or amino acid changes.

- Nucleotide substitution (e.g. 76A>T) - The number is the position of the nucleotide from the 5' end, the first letter represents the wild type nucleotide, and the second letter represents the nucleotide which replaced the wild type. In the given example, the adenine at the 76th position was replaced by a thymine.
- If it becomes necessary to differentiate between mutations in genomic DNA, mitochondrial DNA, and RNA, a simple convention is used. For example, if the 100th base of a nucleotide sequence mutated from G to C, then it would be written as g.100G>C if the mutation occurred in genomic DNA, m.100G>C if the mutation occurred in mitochondrial DNA, or r.100g>c if the mutation occurred in RNA. Note that for mutations in RNA, the nucleotide code is written in lower case.
- Amino acid substitution (e.g. D111E) – The first letter is the one letter code of the wild type amino acid, the number is the position of the amino acid from the N-terminus, and the second letter is the one letter code of the amino acid present in the mutation. Nonsense mutations are represented with an X for the second amino acid (e.g. D111X).
- Amino acid deletion (e.g. ÄF508) – The Greek letter Ä (delta) indicates a deletion. The letter refers to the amino acid present in the wild type and the number is the position from the N terminus of the amino acid were it to be present as in the wild type.

Harmful Mutations

Changes in DNA caused by mutation can cause errors in protein sequence, creating partially or completely non-functional proteins. To function correctly, each cell depends on thousands of proteins to function in the right places at the right times.

When a mutation alters a protein that plays a critical role in the body, a medical condition can result. A condition caused by mutations in one or more genes is called a genetic disorder. Some mutations alter a gene's DNA base sequence but do not change the function of the protein made by the gene.

Studies of the fly Drosophila melanogastersuggest that if a mutation does change a protein, this will probably be harmful, with about 70 percent of these mutations having damaging effects, and the remainder being either neutral or weakly beneficial. However, studies in yeast have shown that only 7% of mutations that are not in genes are harmful.

If a mutation is present in a germ cell, it can give rise to offspring that carries the mutation in all of its cells. This is the case in hereditary diseases. On the other hand, a mutation may occur in asomatic cell of an organism. Such mutations will be present in all descendants of this cell within the same organism, and certain mutations can cause the cell to become malignant, and thus causecancer.

Often, gene mutations that could cause a genetic disorder are repaired by the DNA repair system of the cell. Each cell has a number of pathways through which enzymes recognize and repair mistakes in DNA. Because DNA can be damaged or mutated in many ways, the process of DNA repair is an important way in which the body protects itself from disease.

Beneficial Mutations

Although mutations that change protein sequences are predominantly harmful to an organism; on occasion, the effect can be neutral or positive in a given environment.

In this case, the mutation may enable the mutant organism to withstand particular environmental stresses better than wild-type organisms, or reproduce more quickly. In these cases a mutation will tend to become more common in a population through natural selection.

For example, a specific 32 base pair deletion in human CCR5 (CCR5-Ä32) confers HIV resistance to homozygotes and delays AIDS onset in heterozygotes. The CCR5 mutation is more common in those of European descent.

One possible explanation of the etiology of the relatively high frequency of CCR5-Ä32 in the European population is that it conferred resistance to the bubonic plague in mid-14th century Europe. People with this mutation were more likely to survive infection; thus its frequency in the population increased.

This theory could explain why this mutation is not found in southern Africa, where the bubonic plague never reached. A newer theory suggests that the selective pressure on the CCR5 Delta 32 mutation was caused by smallpox instead of the bubonic plague.

Another example is Sickle cell disease, a blood disorder in which the body produces an abnormal type of the oxygen-carrying substance hemoglobin in the red blood cells. One-third of all indigenousinhabitants of Sub-Saharan Africa carry the gene, because in areas where malaria is common, there is a survival value in carrying only a single sickle-cell gene (sickle cell trait). Those with only one of the two alleles of the sickle-cell disease are more resistant to malaria, since the infestation of the malaria plasmodium is halted by the sickling of the cells which it infests.

Prion Mutation

Prions are proteins and do not contain genetic material. However, prion replication has been shown to be subject to mutation and natural selection just like other forms of replication.

Nucleotide

Nucleotides are molecules that, when joined together, make up the structural units of RNA and DNA. In addition, nucleotides participate in cellular signaling (cyclic guanosine monophosphate andcyclic adenosine monophosphate), and are incorporated into important cofactors of enzymatic reactions (coenzyme A, flavin adenine dinucleotide, flavin mononucleotide, and nicotinamide adenine dinucleotide phosphate). Nucleotide derivatives such as the nucleoside triphosphates play central roles in metabolism, in which capacity they serve as sources of chemical energy (adenosine triphosphate and guanosine triphosphate).

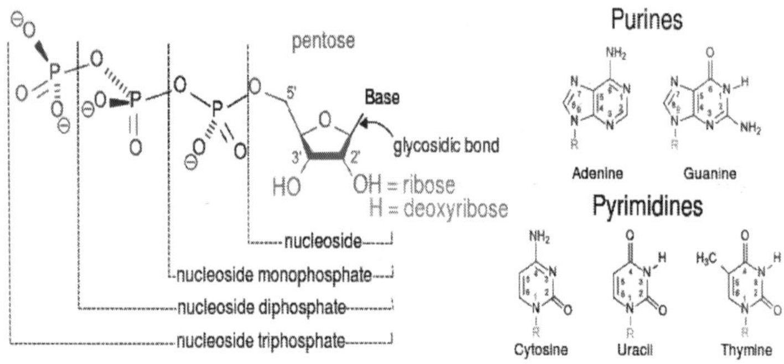

Fig. 7: Structural elements of the most common nucleotides

Nucleotide Structure

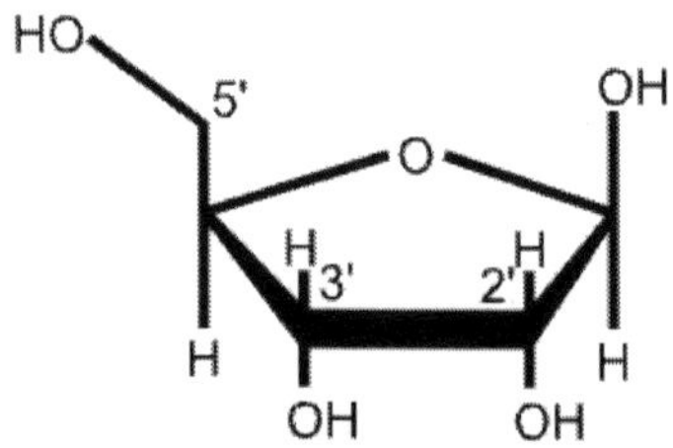

Fig. 8: Ribose structure indicating numbering of carbon atoms

A nucleotide is composed of a nucleobase (nitrogenous base), a five-carbon sugar (either ribose or 2'-deoxyribose), and one phosphate group. Together, the nucleobase and sugar compose a nucleoside. The phosphate groups form bonds with either the 2, 3, or 5-carbon of the sugar, with the 5-carbon site most common. Cyclic nucleotides form when the phosphate group is bound to two of the sugar's hydroxyl groups. Ribonucleotides are nucleotides where the sugar is ribose, and deoxyribonucleotides contain the sugar deoxyribose. Nucleotides can contain either a purine or a pyrimidine base.

Nucleic acids are polymeric macromolecules made from nucleotide monomers. In DNA, the purine bases are adenine and guanine, while the pyrimidines arethymine and cytosine. RNA uses uracil in place of thymine. Adenine always pairs with thymine by 2 hydrogen bonds, while guanine pairs with cytosine through 3 hydrogen bonds, each due to their unique structures.

Synthesis

Nucleotides can be synthesized by a variety of means both in vitro and in vivo.

In vivo, nucleotides can be synthesized de novo or recycled through salvage pathways. The components used in de novo nucleotide synthesis are derived from biosynthetic precursors of carbohydrate and amino acid metabolism, and from ammonia and carbon dioxide. The liver is the major organ of de novo synthesis of all four nucleotides. De novo synthesis of pyrimidines and purines follows two different pathways. Pyrimidines are synthesized first from aspartate and carbamoyl-phosphate in the cytoplasm to the common precursor ring structure orotic acid, onto which a phosphorylated ribosyl unit is covalently linked. Purines, however, are first synthesized from the sugar template onto which the ring synthesis occurs. For reference, the syntheses of the purine and pyrimidine nucleotides are carried out by several enzymes in the cytoplasm of the cell, not within a specific organelle. Nucleotides undergo breakdown such that useful parts can be reused in synthesis reactions to create new nucleotides.

In vitro, protecting groups may be used during laboratory production of nucleotides. A purified nucleoside is protected to create a phosphoramidite, which can then be used to obtain analogues not found in nature and/or to synthesize an oligonucleotide.

Pyrimidine Ribonucleotide Synthesis

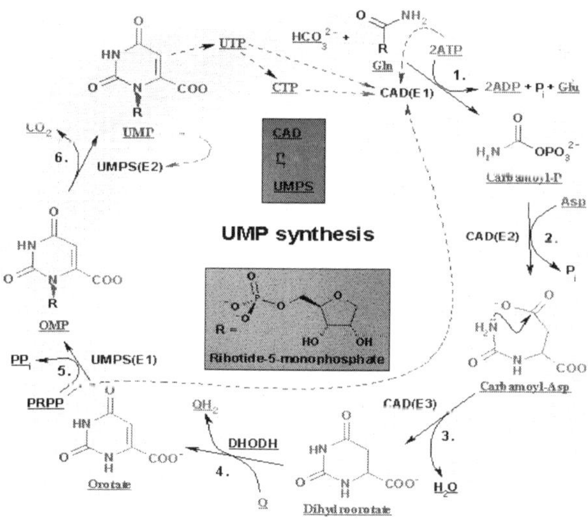

Fig. 9: The synthesis of UMP.

The color scheme is as follows: enzymes, coenzymes, substrate names,inorganic molecules

The synthesis of the pyrimidines CTP and UTP occurs in the cytoplasm and starts with the formation of carbamoyl phosphate from glutamine and CO_2. Next, aspartate undergoes a condensation reaction with carbamoyl-phosphate to form orotic acid.

In a subsequent cyclization reaction, the enzyme Aspartate carbamoyltransferase forms N-carbamoyl-aspartate which is converted into dihydroorotic acid by Dihydroorotase. The latter is converted to orotate by Dihydroorotate oxidase. The net reaction is:

(S)-Dihydroorotate + O_2 = Orotate + H_2O_2

Orotate is covalently linked with a phosphorylated ribosyl unit. The covalent linkage between the ribose and pyrimidine occurs at position C_1 of the ribose unit, which contains a pyrophosphate, and N_1 of the pyrimidine ring. Orotate phosphoribosyltransferase (aka "PRPP transferase") catalyzes the net reaction yielding orotidine monophosphate (OMP):

Orotate + 5-Phospho-á-D-ribose 1-diphosphate (aka. "PRPP") = Orotidine 5'-phosphate + Pyrophosphate

Orotidine-5-phosphate is decarboxylated by Orotidine-5'-phosphate decarboxylase to form uridine monophosphate (UMP). PRPP transferase catalyzes both the ribosylation and decarboxylation reactions, forming UMP from orotic acid in the presence of PRPP.

It is from UMP that other pyrimidine nucleotides are derived. UMP is phosphorylated by two kinases to uridine triphosphate (UTP) via two sequential reactions with ATP. First the diphosphate form UDP is produced, which in turn is phosphorylated to UTP. Both steps are fueled by ATP hydrolysis:

ATP + UMP = ADP + UDP UDP + ATP = UTP + ADP

CTP is subsequently formed by amination of UTP by the catalytic activity of CTP synthetase. Glutamine is the NH_3 donor and the reaction is fueled by ATP hydrolysis, too:

UTP + Glutamine + ATP + H_2O = CTP + ADP + P_i

Cytidine monophosphate (CMP) is derived from cytidine triphosphate (CTP) with subsequent loss of two phosphates.

Purine Ribonucleotide Synthesis

The atoms which are used to build the purine nucleotides come from a variety of sources:

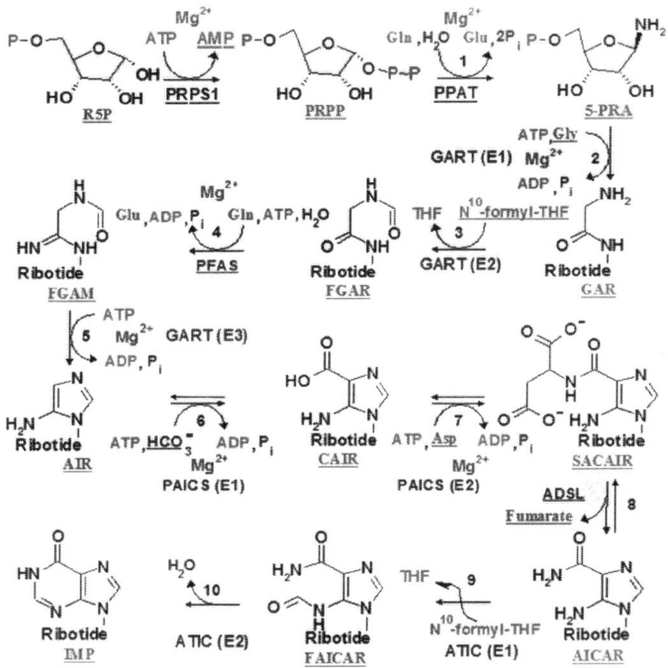

Fig. 10:

The synthesis of IMP. The color scheme is as follows: enzymes,coenzymes, substrate names, metal ions, inorganic molecules

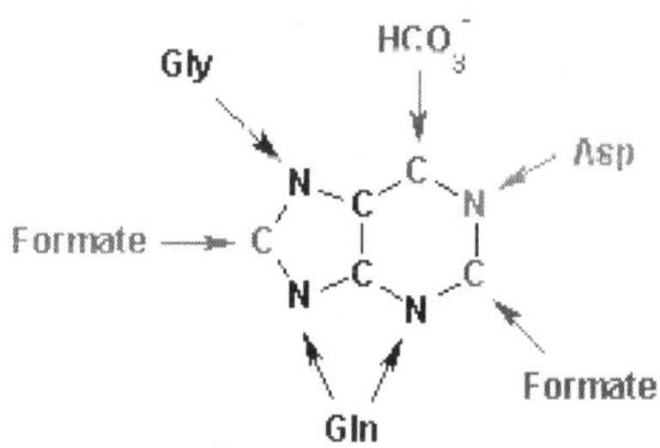

Fig. 11:

The biosynthetic origins of purine ring atoms

N1 arises from the amine group of Asp

C2 and C8 originate from formate

N3 and N9 are contributed by the amide group of Gln

C4, C5 and N7 are derived from Gly

C6 comes from HCO3- (CO2)

The de novo synthesis of purine nucleotides by which these precursors are incorporated into the purine ring proceeds by a 10-step pathway to the branch-point intermediate IMP, the nucleotide of the base hypoxanthine. AMP and GMP are subsequently synthesized from this intermediate via separate, two-step pathways. Thus, purine moieties are initially formed as part of theribonucleotides rather than as free bases.

Six enzymes take part in IMP synthesis. Three of them are multifunctional:

- GART (reactions 2, 3, and 5)
- PAICS (reactions 6, and 7)
- ATIC (reactions 9, and 10)

The pathway starts with the formation of PRPP. PRPS1 is the enzyme that activates R5P, which is formed primarily by thepentose phosphate pathway, to PRPP by reacting it with ATP. The reaction is unusual in that a pyrophosphoryl group is directly transferred from ATP to C_1 of R5P and that the product has the á configuration about C1. This reaction is also shared with the pathways for the synthesis of Trp, His, and the pyrimidine nucleotides. Being on a major metabolic crossroad and requiring much energy, this reaction is highly regulated.

In the first reaction unique to purine nucleotide biosynthesis, PPAT catalyzes the displacement of PRPP's pyrophosphategroup (PP_i) by an amide nitrogen donated from either glutamine (N), glycine (N&C), aspartate (N), folic acid (C_1), or CO_2. This is the committed step in purine synthesis. The reaction occurs with the inversion of configuration about ribose C_1, thereby forming â-5-phosphorybosylamine (5-PRA) and establishing the anomeric form of the future nucleotide.

Next, a glycine is incorporated fueled by ATP hydrolysis and the carboxyl group forms an amine bond to the NH_2 previously introduced. A one-carbon unit from folic acid coenzyme N_{10}-formyl-THF is then added to the amino group of the substituted glycine followed by the closure of the imidazole ring. Next, a second NH_2 group is transferred from a glutamine to the first carbon of the glycine unit. A carboxylation of the second carbon of the glycin unit is concomittantly added. This new carbon is modified by the additional of a third NH_2 unit, this time transferred from an aspartate residue. Finally, a second one-carbon unit from formyl-THF is added to the nitrogen group and the ring covalently closed to form the common purine precursor inosine monophosphate (IMP).

Inosine monophosphate is converted to adenosine monophosphate in two steps. First, GTP hydrolysis fuels the addition of aspartate to IMP by adenylosuccinate synthase, substituting the carbonyl oxygen for a nitrogen and forming the intermediate adenylosuccinate. Fumarate is then cleaved off forming adenosine monophosphate. This step is catalyzed by adenylosuccinate lyase.

Inosine monophosphate is converted to guanosine monophosphate by the oxidation of IMP forming xanthylate, followed by the insertion of an amino group at C_2. NAD is the electron acceptor in the oxidation reaction. The amide group transfer from glutamine is fueled by ATP hydrolysis.

Pyramidine and Purine Degradation

In humans, pyrimidine rings (C, T, U) can be degraded completely to CO_2 and NH_3 (urea excretion). That having been said, purine rings (G, A) cannot. Instead they are degraded to the metabolically inert uric acid which is then excreted from the body. Uric acid is formed when GMP is split into the base guanine and ribose. Guanine is deaminated to xanthine which in turn is oxidized to uric acid. This last reaction is irreversible. Similarly, uric acid can be formed when AMP is deaminated to IMP from which the ribose unit is removed to form hypoxanthine. Hypoxanthine is oxidized to xanthine and finally to uric acid. Instead of uric acid secretion, guanine and IMP can be used for recycling purposes and nucleic acid synthesis in the presence of PRPP and aspartate (NH_3 donor).

Length Unit

Nucleotide (abbreviated nt) is a common length unit for single-stranded RNA, similar to how base pair is a length unit for double-stranded DNA.

Abbreviation Codes for Degenerate Bases

The IUPAC has designated the symbols for nucleotides. Apart from the five (A, G, C, T/U) bases, often degenerate bases are used especially for designing PCR primers. These nucleotide codes are listed here.

IUPAC nucleotide code	*Base*
A	Adenine
C	Cytosine
G	Guanine
T (or U)	Thymine (or Uracil)
R	A or G
Y	C or T (U)
S	G or C
W	A or T (U)
K	G or T (U)
M	A or C
B	C or G or T (U)
D	A or G or T (U)
H	A or C or T (U)
V	A or C or G
N	any base
. or -	gap

Genetic Variation

Genetic variation, variation in alleles of genes, occurs both within and among populations. Genetic variation is important because it provides the "raw material" for natural selection. Genetic variation is brought about by mutation, a change in a chemical structure of a gene. Polyploidy is an example of chromosomal mutation. Polyploidy is a condition wherein organisms have three or more sets of genetic variation (3n or more).

Among Individuals within a Population

Genetic variation among individuals within a population can be identified at a variety of levels. It is possible to identify genetic variation from observations of phenotypic variation in either quantitative traits (traits that vary continuously and are coded for by many genes, e.g., leg length in dogs) or discrete traits (traits that fall into discrete categories and are coded for by one or a few genes (e.g., white, pink, red petal color in certain flowers).

Genetic variation can also be identified by examining variation at the level of enzymes using the process of protein electrophoresis.

Polymorphic genes have more than one allele at each locus. Half of the genes that code for enzymes in insects and plants may be polymorphic, whereas polymorphisms are less common in vertebrates.

Ultimately, genetic variation is caused by variation in the order of bases in the nucleotides in genes. New technology now allows scientists to directly sequence DNA which has identified even more genetic variation than was previously detected by protein electrophoresis. Examination of DNA has shown genetic variation in both coding regions and in the non-coding intron region of genes.

Genetic variation will result in phenotypic variation if variation in the order of nucleotides in the DNA sequence results in a difference in the order of amino acids in proteins coded by that DNA sequence, and if the resultant differences in amino acid sequence influence the shape, and thus the function of the enzyme.

Between Populations

Geographic variation in genes often occurs among populations living in different locations. Geographic variation may be due to differences in selective pressures or to genetic drift.

Measurement

Genetic variation within a population is commonly measured as the percentage of gene loci that are polymorphic or the percentage of gene loci in individuals that are heterozygous.

Sources

Mutations are the ultimate source of genetic variation because they alter the order of bases in the nucleotides of DNA. Mutations are likely to be rare and most mutations are neutral or deleterious, but in some instances the new alleles can be favored by natural selection.

Genetic variation can also be produced by the recombination of chromosomes that occurs during sexual reproduction, called independent assortment.

Crossing over and random segregation during meiosis can result in the production of new alleles or new combinations of alleles.

Variation and recombination can be facilitated by transposable and transposed genetic elements, commonly known as endogenous retroviruses, LINEs, SINEs, etc.

Maintenance in Populations

A variety of factors maintain genetic variation in populations. Potentially harmful recessive alleles can be hidden from selection in the heterozygous individuals in populations of diploid organisms (recessive alleles are only expressed in the less common homozygous individuals). Natural selection can also maintain genetic variation in balanced polymorphisms. Balanced polymorphisms may occur when heterozygotes are favored or when selection is frequency dependent.

Evolution

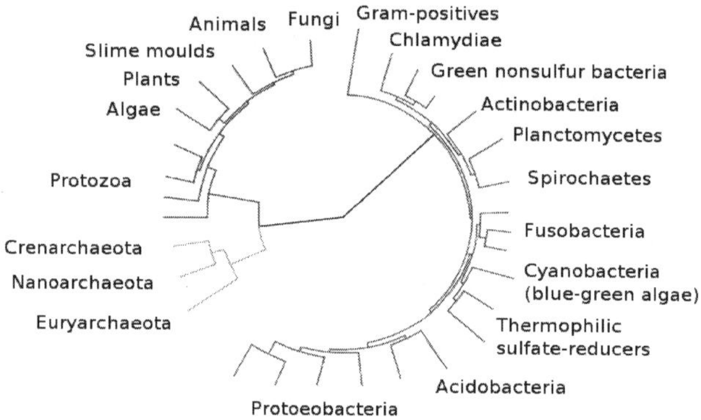

Fig. 12:

Evolution is any change across successive generations in the heritable characteristics of biological populations. Evolutionary processes give rise to diversity at every level of biological organisation, including species, individual organisms and molecules such as DNA and proteins.

Life on Earth originated and then evolved from a universal common ancestor approximately 3.7 billion years ago. Repeated speciation and the divergence of life can be traced through shared sets of biochemical and morphological traits, or by shared DNA sequences. These homologous traits and sequences are more similar among species that share a more recent common ancestor, and can be used to reconstruct evolutionary histories, using both existing species and the fossil record. Existing patterns of biodiversity have been shaped both by speciation and by extinction.

The theory of evolution by natural selection was formulated by Charles Darwin, who was the first to recognise natural selection as an important cause of evolution. Evolution by natural selection is a consequence of three widely accepted premises: 1) more offspring are produced than can possibly survive, 2) traits vary among individuals, leading to differential rates of survival and reproduction, and 3) trait differences are heritable. Thus, when members of a population die, they are replaced by individuals that are not born from random parents. Instead, these new members are born from parents that are betteradapted to the environment in which natural selection took place. This can cause the evolution of traits that are seemingly fitted for the functional roles they perform. Natural selection is the only known cause of adaptation, but not the only known cause of evolution. Other, nonadaptive causes of evolutioninclude mutation and genetic drift.

In the early 20th century, genetics was integrated with Darwin's theory of evolution by natural selection. All other branches of biology were made compatible with the resulting synthesis while previously held notions about evolution, such as orthogenesis and "progress", were made obsolete. Scientists continue to study evolution using both observational data and experiments in both the field and the laboratory. Biologists agree that descent with modification is one of the most reliably established facts in science. Discoveries in evolutionary biology have made a significant impact not just within the traditional branches of biology, but also in other academic disciplines (e.g., anthropology and psychology) and on society at large.

History of Evolutionary Thought

The proposal that one type of animal could descend from an animal of another type goes back to some of the first pre Socratic Greek philosophers, such as Anaximander and Empedocles. In contrast to these materialistic views, Aristotle understood all natural things, not only living things, as being imperfect actualizations of different fixed natural possibilities, known as "forms", "ideas", or (in Latin translations) "species". This was part of his teleological understanding of nature in which all things have an intended role to play in a divine cosmic order. Variations of this idea became the standard understanding of the Middle Ages, and were integrated into Christian learning, but Aristotle did not demand that real types of animals corresponded one-for-one with exact metaphysical forms, and specifically gave examples of how new types of living things could come to be.

In the 17th century the new method of modern science rejected Aristotle's approach, and sought explanations of natural phenomena in terms of laws of nature which were the same for all visible things, and did not need to assume any fixed natural categories, nor any divine cosmic order. But this new approach was slow to take root in the biological sciences, which became the last bastion of the concept of fixed natural types.

John Ray used one of the previously more general terms for fixed natural types, "species", to apply to animal and plant types, but unlike Aristotle he strictly identified each type of living thing as a species, and proposed that each species can be defined by the features that perpetuate themselves each generation. These species were designed by God, but showing differences caused by local conditions. The biological classification introduced by Carolus Linnaeus in 1735 also viewed species as fixed according to a divine plan.

Other naturalists of this time speculated on evolutionary change of species over time according to natural laws. Maupertuis wrote in 1751 of natural modifications occurring during reproduction and accumulating over many generations to produce new species.

Buffon suggested that species could degenerate into different organisms, and Erasmus Darwin proposed that all warm-blooded animals could have descended from a single micro-organism (or "filament").

The first fully-fledged evolutionary scheme was Lamarck's "transmutation" theory of 1809 which envisaged spontaneous generation continually producing simple forms of life developed greater complexity in parallel lineages with an inherent progressive tendency, and that on a local level these lineages adapted to the environment by inheriting changes caused by use or disuse in parents. (The latter process was later called Lamarckism.)

These ideas were condemned by establishment naturalists as speculation lacking empirical support. In particular Georges Cuvier insisted that species were unrelated and fixed, their similarities reflecting divine design for functional needs.

In the meantime, Ray's ideas of benevolent design had been developed by William Paley into a natural theology which proposed complex adaptations as evidence of divine design, and was admired by Charles Darwin.

Fig. 13:

In 1842 Charles Darwin penned his first sketch of what became On the Origin of Species.

The critical break of biology from the concept of fixed species is generally seen as the theory of evolution of Charles Darwin. Partly influenced by An Essay on the Principle of Population by Thomas Robert Malthus, Darwin noted that population growth would lead to a "struggle for existence" where favorable variations could prevail as others perished. Each generation, many offspring fail to survive to an age of reproduction because of limited resources.

This could explain the diversity of animals and plants from a common ancestry through the working of natural laws working the same for all types of thing. Darwin was developing his theory of "natural selection" from 1838 onwards until Alfred Russel Wallace sent him a similar theory in 1858. Both men presented theirseparate papers to the Linnean Society of London. At the end of 1859, Darwin's publication of On the Origin of Species explained natural selection in detail and in a way that lead to an increasingly wide acceptance of Darwinian evolution. Thomas Henry Huxley applied Darwin's ideas to humans, usingpaleontology and comparative anatomy to provide strong evidence that humans and apes shared a common ancestry. Some were disturbed by this since it implied that humans did not have a special place in the universe.

Precise mechanisms of reproductive heritability and the origin of new traits remained a mystery. Towards this end, Darwin developed his provisional theory ofpangenesis. In 1865 Gregor Mendel reported that traits were inherited in a predictable manner through the independent assortment and segregation of elements (later known as genes). Mendel's laws of inheritance eventually supplanted most of Darwin's pangenesis theory. August Weismann made the important distinction between germ cells (sperm and eggs) and somatic cells of the body, demonstrating that heredity passes through the germ line only.Hugo de Vries connected Darwin's pangenesis theory to Wiesman's germ/soma cell distinction and proposed that Darwin's pangenes were concentrated in the cell nucleus and when expressed they could move into the cytoplasm to change the cells structure. De Vries was also one of the researchers who made Mendel's work well-known, believing that Mendelian traits corresponded to the transfer of heritable variations along the germline. To explain how new variants originate, De Vries developed a mutation theory that led to a temporary rift between those who accepted Darwinian evolution and biometricians who allied with de Vries. At the turn of the 20th century, pioneers in the field of population genetics, such as J.B.S. Haldane, Sewall Wright, and Ronald Fisher, set the foundations of evolution onto a robust statistical philosophy. The false contradiction between Darwin's theory, genetic mutations, and Mendelian inheritance was thus reconciled.

In the 1920s and 1930s a modern evolutionary synthesis connected natural selection, mutation theory, and Mendelian inheritance into a unified theory that applied generally to any branch of biology. The modern synthesis was able to explain patterns observed across species in populations, through fossil transitions in palaeontology, and even complex cellular mechanisms in developmental biology. The publication of the structure of DNA by James Watson and Francis Crick in 1953 demonstrated a physical basis for inheritance. Molecular biology improved our understanding of the relationship between genotype and phenotype. Advancements were also made in phylogenetic systematics, mapping the transition of traits into a comparative and testable framework through the publication and use of evolutionary trees. In 1973, evolutionary biologist Theodosius Dobzhansky penned that "nothing in biology makes sense except in the light of evolution", because it has brought to light the relations of what first seemed disjointed facts in natural history into a coherent explanatory body of knowledge that describes and predicts many observable facts about life on this planet.

Since then, the modern synthesis has been further extended to explain biological phenomena across the full and integrative scale of the biological hierarchy, from genes to species. This extension has been dubbed "eco-evo-devo".

Heredity

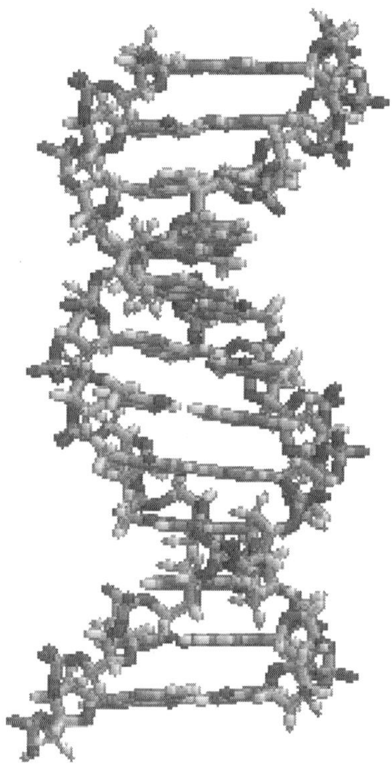

Fig. 14: DNA structure. Bases are in the centre, surrounded by phosphate–sugar chains in adouble helix.

Evolution in organisms occurs through changes in heritable traits – particular characteristics of an organism. In humans, for example, eye colour is an inherited characteristic and an individual might inherit the "brown-eye trait" from one of their parents. Inherited traits are controlled by genes and the complete set of genes within an organism's genome is called its genotype.

The complete set of observable traits that make up the structure and behaviour of an organism is called its phenotype. These traits come from the interaction of its genotype with the environment. As a result, many aspects of an organism's phenotype are not inherited. For example, suntanned skin comes from the interaction between a person's genotype and sunlight; thus, suntans are not passed on to people's children. However, some people tan more easily than others, due to differences in their genotype; a striking example are people with the inherited trait of albinism, who do not tan at all and are very sensitive to sunburn.

Heritable traits are known to be passed from one generation to the next via DNA, a molecule that encodes genetic information. DNA is a long polymer composed of four types of bases. The sequence of bases along a particular DNA molecule specify the genetic information, in a manner similar to a sequence of letters spelling out a sentence. Before a cell divides, the DNA is copied, so that each of the resulting two cells will inherit the DNA sequence. Portions of a DNA molecule that specify a single functional unit are called genes; different genes have different sequences of bases. Within cells, the long strands of DNA form condensed structures called chromosomes. The specific location of a DNA sequence within a chromosome is known as a locus. If the DNA sequence at a locus varies between individuals, the different forms of this sequence are called alleles. DNA sequences can change through mutations, producing new alleles. If a mutation occurs within a gene, the new allele may affect the trait that the gene controls, altering the phenotype of the organism. However, while this simple correspondence between an allele and a trait works in some cases, most traits are more complex and are controlled by multiple interacting genes.

Recent findings have confirmed important examples of heritable changes that cannot be explained by changes to the sequence of nucleotides in the DNA. These phenomena are classed as epigenetic inheritance systems. DNA methylation marking chromatin, self-sustaining metabolic loops, gene silencing by RNA interferenceand the three dimensional conformation of proteins (such as prions) are areas where epigenetic inheritance systems have been discovered at the organismic level.Developmental biologists suggest that complex interactions in genetic networks and communication among cells can lead to heritable variations that may underlay some of the mechanics in developmental plasticity and canalization. Heritability may also occur at even larger scales. For example, ecological inheritance through the process of niche construction is defined by the regular and repeated activities of organisms in their environment. This generates a legacy of effects that modify and feed back into the selection regime of subsequent generations. Descendants inherit genes plus environmental characteristics generated by the ecological actions of ancestors. Other examples of heritability in evolution that are not under the direct control of genes include the inheritance of cultural traits and symbiogenesis.

Variation

Fig. 15: White peppered moth

Fig. 16: Black morph in peppered moth evolution

An individual organism's phenotype results from both its genotype and the influence from the environment it has lived in. A substantial part of the variation in phenotypes in a population is caused by the differences between their genotypes. The modern evolutionary synthesis defines evolution as the change over time in this genetic variation. The frequency of one particular allele will become more or less prevalent relative to other forms of that gene. Variation disappears when a new allele reaches the point of fixation — when it either disappears from the population or replaces the ancestral allele entirely.

Natural selection will only cause evolution if there is enough genetic variation in a population. Before the discovery of Mendelian genetics, one common hypothesis was blending inheritance. But with blending inheritance, genetic variance would be rapidly lost, making evolution by natural selection implausible.

The Hardy-Weinberg principle provides the solution to how variation is maintained in a population with Mendelian inheritance. The frequencies of alleles (variations in a gene) will remain constant in the absence of selection, mutation, migration and genetic drift.

Variation comes from mutations in genetic material, reshuffling of genes through sexual reproduction and migration between populations (gene flow). Despite the constant introduction of new variation through mutation and gene flow, most of the genome of a species is identical in all individuals of that species.

However, even relatively small differences in genotype can lead to dramatic differences in phenotype: for example, chimpanzees and humans differ in only about 5% of their genomes.

Mutation

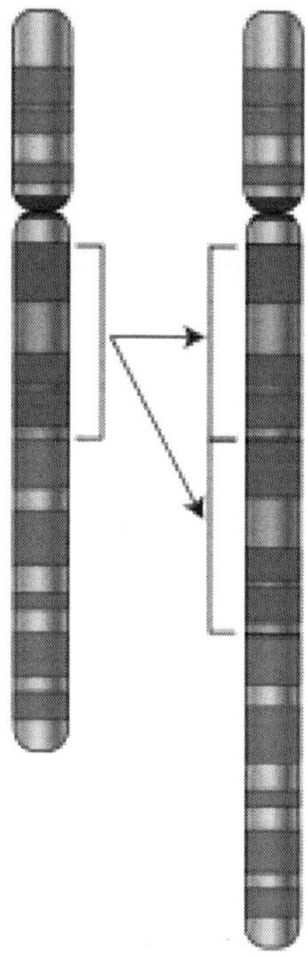

Fig. 17: Duplication of part of achromosome.

Mutations are changes in the DNA sequence of a cell's genome.

When mutations occur, they can either have no effect, alter the product of a gene, or prevent the gene from functioning. Based on studies in the fly Drosophila melanogaster, it has been suggested that if a mutation changes a protein produced by a gene, this will probably be harmful, with about 70% of these mutations having damaging effects, and the remainder being either neutral or weakly beneficial.

Mutations can involve large sections of a chromosome becoming duplicated (usually by genetic recombination), which can introduce extra copies of a gene into a genome.Extra copies of genes are a major source of the raw material needed for new genes to evolve. This is important because most new genes evolve within gene families from pre-existing genes that share common ancestors. For example, the human eye uses four genes to make structures that sense light: three for colour vision and one for night vision; all four are descended from a single ancestral gene.

New genes can be generated from an ancestral gene when a duplicate copy mutates and acquires a new function. This process is easier once a gene has been duplicated because it increases the redundancy of the system; one gene in the pair can acquire a new function while the other copy continues to perform its original function. Other types of mutations can even generate entirely new genes from previously noncoding DNA.

The generation of new genes can also involve small parts of several genes being duplicated, with these fragments then recombining to form new combinations with new functions. When new genes are assembled from shuffling pre-existing parts, domains act as modules with simple independent functions, which can be mixed together to produce new combinations with new and complex functions. For example, polyketide synthases are large enzymes that make antibiotics; they contain up to one hundred independent domains that each catalyze one step in the overall process, like a step in an assembly line.

Chapter 5

Sex and Recombination

In asexual organisms, genes are inherited together, or linked, as they cannot mix with genes of other organisms during reproduction. In contrast, the offspring of sexual organisms contain random mixtures of their parents' chromosomes that are produced through independent assortment.

In a related process called homologous recombination, sexual organisms exchange DNA between two matching chromosomes. Recombination and reassortment do not alter allele frequencies, but instead change which alleles are associated with each other, producing offspring with new combinations of alleles. Sex usually increases genetic variation and may increase the rate of evolution.

Gene Flow

Gene flow is the exchange of genes between populations and between species. It can therefore be a source of variation that is new to a population or to a species. Gene flow can be caused by the movement of individuals between separate populations of organisms, as might be caused by the movement of mice between inland and coastal populations, or the movement of pollen between heavy metal tolerant and heavy metal sensitive populations of grasses.

Gene transfer between species includes the formation of hybrid organisms and horizontal gene transfer. Horizontal gene transfer is the transfer of genetic material from one organism to another organism that is not its offspring; this is most common among bacteria.

In medicine, this contributes to the spread of antibiotic resistance, as when one bacteria acquires resistance genes it can rapidly transfer them to other species. Horizontal transfer of genes from bacteria to eukaryotes such as the yeast Saccharomyces cerevisiae and the adzuki bean beetle Callosobruchus chinensishas occurred. An example of larger-scale transfers are the eukaryotic bdelloid rotifers, which have received a range of genes from bacteria, fungi and plants.

Viruses can also carry DNA between organisms, allowing transfer of genes even across biological domains.

Large-scale gene transfer has also occurred between the ancestors of eukaryotic cells and bacteria, during the acquisition of chloroplasts and mitochondria. It is possible that eukaryotes themselves originated from horizontal gene transfers between bacteria and archaea.

Mechanisms

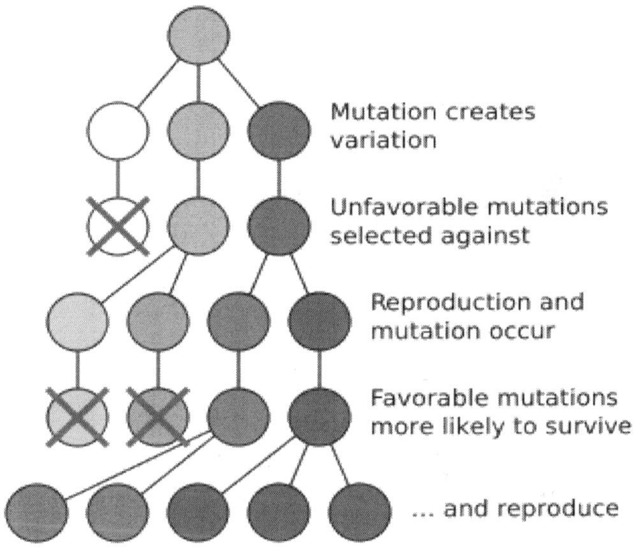

Fig. 1: Mutation followed by natural selection, results in a population with darker colouration.

From a Neo-Darwinian perspective, evolution occurs when there are changes in the frequencies of alleles within a population of interbreeding organisms. For example, the allele for black colour in a population of moths becoming more common. Mechanisms that can lead to changes in allele frequencies include natural selection, genetic drift, genetic hitchhiking, mutation and gene flow.

Natural Selection

Evolution by means of natural selection is the process by which genetic mutations that enhance reproduction become and remain, more common in successive generations of a population. It has often been called a "self-evident" mechanism because it necessarily follows from three simple facts:

- Heritable variation exists within populations of organisms.
- Organisms produce more offspring than can survive.
- These offspring vary in their ability to survive and reproduce.

These conditions produce competition between organisms for survival and reproduction. Consequently, organisms with traits that give them an advantage over their competitors pass these advantageous traits on, while traits that do not confer an advantage are not passed on to the next generation.

The central concept of natural selection is the evolutionary fitness of an organism. Fitness is measured by an organism's ability to survive and reproduce, which determines the size of its genetic contribution to the next generation.

However, fitness is not the same as the total number of offspring: instead fitness is indicated by the proportion of subsequent generations that carry an organism's genes.

For example, if an organism could survive well and reproduce rapidly, but its offspring were all too small and weak to survive, this organism would make little genetic contribution to future generations and would thus have low fitness.

If an allele increases fitness more than the other alleles of that gene, then with each generation this allele will become more common within the population.

These traits are said to be "selected for". Examples of traits that can increase fitness are enhanced survival and increased fecundity. Conversely, the lower fitness caused by having a less beneficial or deleterious allele results in this allele becoming rarer — they are "selected against".

Importantly, the fitness of an allele is not a fixed characteristic; if the environment changes, previously neutral or harmful traits may become beneficial and previously beneficial traits become harmful.

However, even if the direction of selection does reverse in this way, traits that were lost in the past may not re-evolve in an identical form.

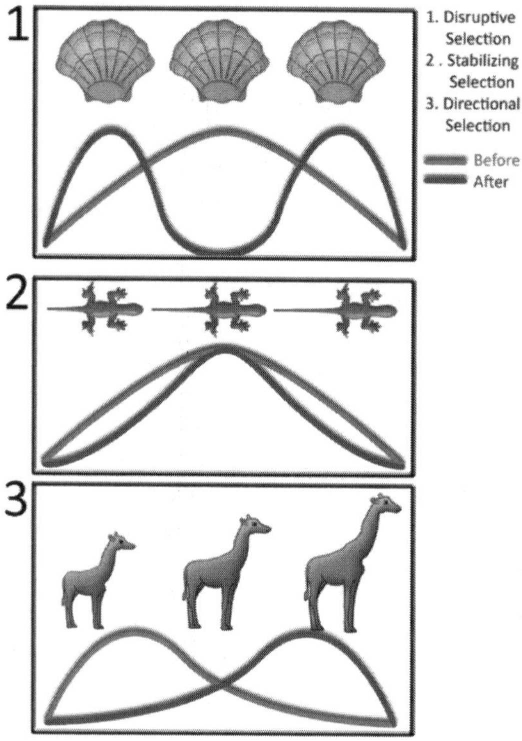

Fig. 2: A chart showing three types of selection. 1. Disruptive selection 2. Stabilizing selection3. Directional selection

Natural selection within a population for a trait that can vary across a range of values, such as height, can be categorised into three different types. The first isdirectional selection, which is a shift in the average value of a trait over time — for example, organisms slowly getting taller. Secondly, disruptive selectionis selection for extreme trait values and often results in two different values becoming most common, with selection against the average value. This would be when either short or tall organisms had an advantage, but not those of medium height. Finally, in stabilizing selection there is selection against extreme trait values on both ends, which causes a decrease in variance around the average value and less diversity. This would, for example, cause organisms to slowly become all the same height.

A special case of natural selection is sexual selection, which is selection for any trait that increases mating success by increasing the attractiveness of an organism to potential mates. Traits that evolved through sexual selection are particularly prominent in males of some animal species, despite traits such as cumbersome antlers, mating calls or bright colours that attract predators, decreasing the survival of individual males. This survival disadvantage is balanced by higher reproductive success in males that show these hard to fake, sexually selected traits. Natural selection most generally makes nature the measure against which individuals and individual traits, are more or less likely to survive. "Nature" in this sense refers to an ecosystem, that is, a system in which organisms interact with every other element, physical as well as biological, in their localenvironment.

Eugene Odum, a founder of ecology, defined an ecosystem as: "Any unit that includes all of the organisms...in a given area interacting with the physical environment so that a flow of energy leads to clearly defined trophic structure, biotic diversity and material cycles (ie: exchange of materials between living and nonliving parts) within the system." Each population within an ecosystem occupies a distinct niche, or position, with distinct relationships to other parts of the system. These relationships involve the life history of the organism, its position in the food chain and its geographic range. This broad understanding of nature enables scientists to delineate specific forces which, together, comprise natural selection.

Natural selection can act at different levels of organisation, such as genes, cells, individual organisms, groups of organisms and species. Selection can act at multiple levels simultaneously. An example of selection occurring below the level of the individual organism are genes called transposons, which can replicate and spread throughout a genome. Selection at a level above the individual, such as group selection, may allow the evolution of co-operation.

Biased Mutation

In addition to being a major source of variation, mutation may also function as a mechanism of evolution when there are different probabilities at the molecular level for different mutations to occur, a process known as mutation bias. If two genotypes, for example one with the nucleotide G and another with the nucleotide A in the same position, have the same fitness, but mutation from G to A happens more often than mutation from A to G, then genotypes with A will tend to evolve. Different insertion vs. deletion mutation biases in different taxa can lead to the evolution of different genome sizes. Developmental or mutational biases have also been observed in morphological evolution. For example, according to the phenotype-first theory of evolution, mutations can eventually cause the genetic assimilation of traits that were previously induced by the environment.

Mutation bias effects are superimposed on other processes. If selection would favor either one out of two mutations, but there is no extra advantage to having both, then the mutation that occurs the most frequently is the one that is most likely to become fixed in a population. Mutations leading to the loss of function of a gene are much more common than mutations that produce a new, fully functional gene. Most loss of function mutations are selected against. But when selection is weak, mutation bias towards loss of function can affect evolution. For example, pigments are no longer useful when animals live in the darkness of caves, and tend to be lost. This kind of loss of function can occur because of mutation bias, and/or because the function had a cost, and once the benefit of the function disappeared, natural selection leads to the loss. Loss of sporulation ability in a bacterium during laboratory evolution appears to have been caused by mutation bias, rather than natural selection against the cost of maintaining sporulation ability. When there is no selection for loss of function, the speed at which loss evolves depends more on the mutation rate than it does on the effective population size, indicating that it is driven more by mutation bias than by genetic drift.

Genetic Drift

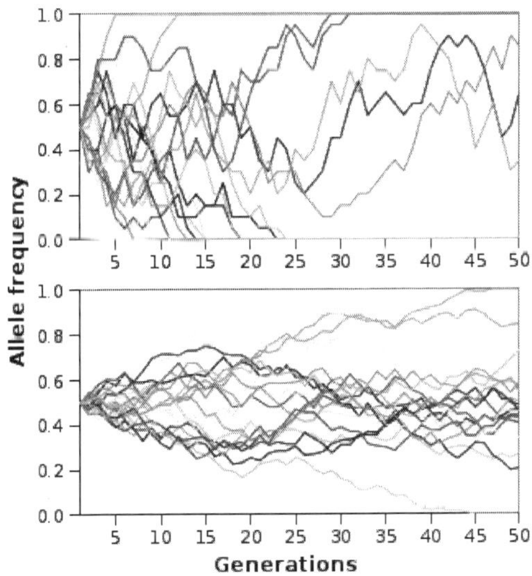

Fig. 3:

Simulation of genetic drift of 20 unlinked alleles in populations of 10 (top) and 100 (bottom). Drift to fixation is more rapid in the smaller population.

Genetic drift is the change in allele frequency from one generation to the next that occurs because alleles are subject to sampling error. As a result, when selective forces are absent or relatively weak, allele frequencies tend to "drift" upward or downward randomly (in a random walk). This drift halts when an allele eventually becomes fixed, either by disappearing from the population, or replacing the other alleles entirely. Genetic drift may therefore eliminate some alleles from a population due to chance alone. Even in the absence of selective forces, genetic drift can cause two separate populations that began with the same genetic structure to drift apart into two divergent populations with different sets of alleles.

It is usually difficult to measure the relative importance of selection and neutral processes, including drift. The comparative importance of adaptive and non-adaptive forces in driving evolutionary change is an area of current research.

The neutral theory of molecular evolution proposed that most evolutionary changes are the result of the fixation of neutral mutations by genetic drift. Hence, in this model, most genetic changes in a population are the result of constant mutation pressure and genetic drift. This form of the neutral theory is now largely abandoned, since it does not seem to fit the genetic variation seen in nature. However, a more recent and better-supported version of this model is the nearly neutral theory, where a mutation that would be neutral in a small population is not necessarily neutral in a large population. Other alternative theories propose that genetic drift is dwarfed by other stochastic forces in evolution, such as genetic hitchhiking, also known as genetic draft.

The time for a neutral allele to become fixed by genetic drift depends on population size, with fixation occurring more rapidly in smaller populations. The number of individuals in a population is not critical, but instead a measure known as the effective population size. The effective population is usually smaller than the total population since it takes into account factors such as the level of inbreeding and the stage of the lifecycle in which the population is the smallest. The effective population size may not be the same for every gene in the same population.

Genetic Hitchhiking

Recombination allows alleles on the same strand of DNA to become separated. However, the rate of recombination is low (approximately two events per chromosome per generation). As a result, genes close together on a chromosome may not always be shuffled away from each other and genes that are close together tend to be inherited together, a phenomenon known as linkage. This tendency is measured by finding how often two alleles occur together on a single chromosome compared to expectations, which is called their linkage disequilibrium. A set of alleles that is usually inherited in a group is called a haplotype. This can be important when one allele in a particular haplotype is strongly beneficial: natural selection can drive a selective sweep that will also cause the other alleles in the haplotype to become more common in the population; this effect is called genetic hitchhiking or genetic draft. Genetic draft caused by the fact that some neutral genes are genetically linked to others that are under selection can be partially captured by an appropriate effective population size.

Gene Flow

Gene flow is the exchange of genes between populations and between species. The presence or absence of gene flow fundamentally changes the course of evolution. Due to the complexity of organisms, any two completely isolated populations will eventually evolve genetic incompatibilities through neutral processes, as in the Bateson-Dobzhansky-Muller model, even if both populations remain essentially identical in terms of their adaptation to the environment.

If genetic differentiation between populations develops, gene flow between populations can introduce traits or alleles which are disadvantageous in the local population and this may lead to organism within these populations to evolve mechanisms that prevent mating with genetically distant populations, eventually resulting in the appearance of new species. Thus, exchange of genetic information between individuals is fundamentally important for the development of the biological species concept (BSC).

During the development of the modern synthesis, Sewall Wright's developed his shifting balance theory that gene flow between partially isolated populations was an important aspect of adaptive evolution. However, recently there has been substantial criticism of the importance of the shifting balance theory.

Outcomes

Evolution influences every aspect of the form and behaviour of organisms. Most prominent are the specific behavioural and physical adaptations that are the outcome of natural selection. These adaptations increase fitness by aiding activities such as finding food, avoiding predators or attracting mates. Organisms can also respond to selection by co-operating with each other, usually by aiding their relatives or engaging in mutually beneficial symbiosis. In the longer term, evolution produces new species through splitting ancestral populations of organisms into new groups that cannot or will not interbreed.

These outcomes of evolution are sometimes divided into macroevolution, which is evolution that occurs at or above the level of species, such as extinction and speciation and microevolution, which is smaller evolutionary changes, such as adaptations, within a species or population. In general, macroevolution is regarded as the outcome of long periods of microevolution. Thus, the distinction between micro- and macroevolution is not a fundamental one – the difference is simply the time involved. However, in macroevolution, the traits of the entire species may be important.

For instance, a large amount of variation among individuals allows a species to rapidly adapt to new habitats, lessening the chance of it going extinct, while a wide geographic range increases the chance of speciation, by making it more likely that part of the population will become isolated. In this sense, microevolution and macroevolution might involve selection at different levels – with microevolution acting on genes and organisms, versus macroevolutionary processes such as species selection acting on entire species and affecting their rates of speciation and extinction.

A common misconception is that evolution has goals or long-term plans; realistically however, evolution has no long-term goal and does not necessarily produce greater complexity. Althoughcomplex species have evolved, they occur as a side effect of the overall number of organisms increasing and simple forms of life still remain more common in the biosphere. For example, the overwhelming majority of species are microscopic prokaryotes, which form about half the world's biomass despite their small size, and constitute the vast majority of Earth's biodiversity. Simple organisms have therefore been the dominant form of life on Earth throughout its history and continue to be the main form of life up to the present day, with complex life only appearing more diverse because it is more noticeable. Indeed, the evolution of microorganisms is particularly important to modern evolutionary research, since their rapid reproduction allows the study of experimental evolution and the observation of evolution and adaptation in real time.

Adaptation

Adaptation is the process that makes organisms better suited to their habitat. Also, the term adaptation may refer to a trait that is important for an organism's survival. For example, the adaptation of horses' teeth to the grinding of grass. By using the term adaptation for the evolutionary process and adaptive trait for the product (the bodily part or function), the two senses of the word may be distinguished. Adaptations are produced by natural selection. The following definitions are due to Theodosius Dobzhansky.

1. Adaptation is the evolutionary process whereby an organism becomes better able to live in its habitat or habitats.
2. Adaptedness is the state of being adapted: the degree to which an organism is able to live and reproduce in a given set of habitats.
3. An adaptive trait is an aspect of the developmental pattern of the organism which enables or enhances the probability of that organism surviving and reproducing.

Adaptation may cause either the gain of a new feature, or the loss of an ancestral feature. An example that shows both types of change is bacterial adaptation to antibiotic selection, with genetic changes causing antibiotic resistance by both modifying the target of the drug, or increasing the activity of transporters that pump the drug out of the cell. Other striking examples are the bacteriaEscherichia coli evolving the ability to use citric acid as a nutrient in a long-term laboratory experiment, Flavobacterium evolving a novel enzyme that allows these bacteria to grow on the by-products of nylon manufacturing, and the soil bacterium Sphingobium evolving an entirely new metabolic pathway that degrades the synthetic pesticide pentachlorophenol. An interesting but still controversial idea is that some adaptations might increase the ability of organisms to generate genetic diversity and adapt by natural selection (increasing organisms'evolvability).

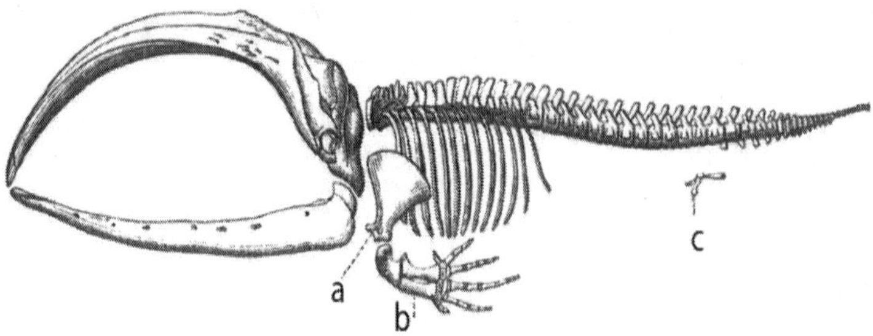

Fig. 4: A baleen whale skeleton, a and b label flipper bones, which wereadapted from front leg bones: while c indicates vestigial leg bones, suggesting an adaptation from land to sea.

Adaptation occurs through the gradual modification of existing structures. Consequently, structures with similar internal organisation may have different functions in related organisms. This is the result of a single ancestral structure being adapted to function in different ways. The bones within bat wings, for example, are very similar to those in mice feet and primate hands, due to the descent of all these structures from a common mammalian ancestor. However, since all living organisms are related to some extent, even organs that appear to have little or no structural similarity, such as arthropod, squid and vertebrate eyes, or the limbs and wings of arthropods and vertebrates, can depend on a common set of homologous genes that control their assembly and function; this is called deep homology.

During evolution, some structures may lose their original function and become vestigial structures. Such structures may have little or no function in a current species, yet have a clear function in ancestral species, or other closely related species. Examples includepseudogenes, the non-functional remains of eyes in blind cave-dwelling fish, wings in flightless birds, and the presence of hip bones in whales and snakes.

However, many traits that appear to be simple adaptations are in fact exaptations: structures originally adapted for one function, but which coincidentally became somewhat useful for some other function in the process. One example is the African lizard Holaspis guentheri, which developed an extremely flat head for hiding in crevices, as can be seen by looking at its near relatives. However, in this species, the head has become so flattened that it assists in gliding from tree to tree—an exaptation. Within cells, molecular machines such as the bacterial flagella and protein sorting machinery evolved by the recruitment of several pre-existing proteins that previously had different functions. Another example is the recruitment of enzymes from glycolysis and xenobiotic metabolism to serve as structural proteins called crystallins within the lenses of organisms' eyes.

A critical principle of ecology is that of competitive exclusion: no two species can occupy the same niche in the same environment for a long time. Consequently, natural selection will tend to force species to adapt to different ecological niches. This may mean that, for example, two species of cichlid fish adapt to live in different habitats, which will minimise the competition between them for food.

An area of current investigation in evolutionary developmental biology is the developmental basis of adaptations and exaptations. This research addresses the origin and evolution of embryonic development and how modifications of development and developmental processes produce novel features. These studies have shown that evolution can alter development to produce new structures, such as embryonic bone structures that develop into the jaw in other animals instead forming part of the middle ear in mammals. It is also possible for structures that have been lost in evolution to reappear due to changes in developmental genes, such as a mutation in chickens causing embryos to grow teeth similar to those of crocodiles. It is now becoming clear that most alterations in the form of organisms are due to changes in a small set of conserved genes.

Co-evolution

Fig. 5: Common Garter Snake (Thamnophis sirtalis sirtalis) which has evolved resistance to tetrodotoxin in its amphibian prey.

Interactions between organisms can produce both conflict and co-operation. When the interaction is between pairs of species, such as a pathogen and ahost, or a predator and its prey, these species can develop matched sets of adaptations. Here, the evolution of one species causes adaptations in a second species. These changes in the second species then, in turn, cause new adaptations in the first species. This cycle of selection and response is called co-evolution. An example is the production of tetrodotoxin in the rough-skinned newt and the evolution of tetrodotoxin resistance in its predator, the common garter snake. In this predator-prey pair, an evolutionary arms race has produced high levels of toxin in the newt and correspondingly high levels of toxin resistance in the snake.

Co-operation

Not all co-evolved interactions between species involve conflict. Many cases of mutually beneficial interactions have evolved. For instance, an extreme cooperation exists between plants and the mycorrhizal fungi that grow on their roots and aid the plant in absorbing nutrients from the soil. This is areciprocal relationship as the plants provide the fungi with sugars from photosynthesis. Here, the fungi actually grow inside plant cells, allowing them to exchange nutrients with their hosts, while sending signals that suppress the plant immune system.

Coalitions between organisms of the same species have also evolved. An extreme case is the eusociality found in social insects, such as bees, termites and ants, where sterile insects feed and guard the small number of organisms in a colony that are able to reproduce. On an even smaller scale, the somatic cells that make up the body of an animal limit their reproduction so they can maintain a stable organism, which then supports a small number of the animal's germ cells to produce offspring. Here, somatic cells respond to specific signals that instruct them whether to grow, remain as they are, or die. If cells ignore these signals and multiply inappropriately, their uncontrolled growth causes cancer.

Such cooperation within species may have evolved through the process of kin selection, which is where one organism acts to help raise a relative's offspring. This activity is selected for because if the helping individual contains alleles which promote the helping activity, it is likely that its kin will also contain these alleles and thus those alleles will be passed on. Other processes that may promote cooperation include group selection, where cooperation provides benefits to a group of organisms.

Speciation

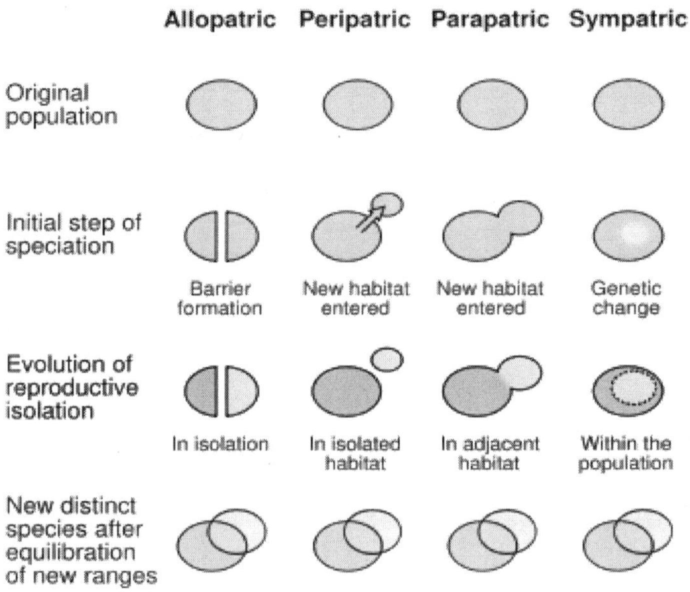

Fig. 6: The four mechanisms of speciation.

Speciation is the process where a species diverges into two or more descendant species.

There are multiple ways to define the concept of "species". The choice of definition is dependent on the particularities of the species concerned. For example, some species concepts apply more readily toward sexually reproducing organisms while others lend themselves better toward asexual organisms. Despite the diversity of various species concepts, these various concepts can be placed into one of three broad philosophical approaches: interbreeding, ecological and phylogenetic. The biological species concept (BSC) is a classic example of the interbreeding approach. Defined by Ernst Mayr in 1942, the BSC states that "species are groups of actually or potentially interbreeding natural populations, which are reproductively isolated from other such groups". Despite its wide and long-term use, the BSC like others is not without controversy, for example because these concepts cannot be applied to prokaryotes, and this is called the species problem. Some researchers have attempted a unifying monistic definition of species, while others adopt a pluralistic approach and suggest that there may be a different ways to logically interpret the definition of a species. ”

Barriers to reproduction between two diverging sexual populations are required for the populations to become new species. Gene flow may slow this process by spreading the new genetic variants also to the other populations. Depending on how far two species have diverged since their most recent common ancestor, it may still be possible for them to produce offspring, as with horses and donkeysmating to produce mules. Such hybrids are generally infertile. In this case, closely related species may regularly interbreed, but hybrids will be selected against and the species will remain distinct. However, viable hybrids are occasionally formed and these new species can either have properties intermediate between their parent species, or possess a totally new phenotype. The importance of hybridisation in producing new species of animals is unclear, although cases have been seen in many types of animals, with thegray tree frog being a particularly well-studied example.

Speciation has been observed multiple times under both controlled laboratory conditions and in nature. In sexually reproducing organisms, speciation results from reproductive isolation followed by genealogical divergence. There are four mechanisms for speciation. The most common in animals is allopatric speciation, which occurs in populations initially isolated geographically, such as byhabitat fragmentation or migration. Selection under these conditions can produce very rapid changes in the appearance and behaviour of organisms. As selection and drift act independently on populations isolated from the rest of their species, separation may eventually produce organisms that cannot interbreed.

The second mechanism of speciation is peripatric speciation, which occurs when small populations of organisms become isolated in a new environment. This differs from allopatric speciation in that the isolated populations are numerically much smaller than the parental population. Here, the founder effect causes rapid speciation after an increase in inbreeding increases selection on homozygotes, leading to rapid genetic change.

The third mechanism of speciation is parapatric speciation. This is similar to peripatric speciation in that a small population enters a new habitat, but differs in that there is no physical separation between these two populations. Instead, speciation results from the evolution of mechanisms that reduce gene flow between the two populations. Generally this occurs when there has been a drastic change in the environment within the parental species' habitat. One example is the grass Anthoxanthum odoratum, which can undergo parapatric speciation in response to localised metal pollution from mines. Here, plants evolve that have resistance to high levels of metals in the soil. Selection against interbreeding with the metal-sensitive parental population produced a gradual change in the flowering time of the metal-resistant plants, which eventually produced complete reproductive isolation. Selection against hybrids between the two populations may cause reinforcement, which is the evolution of traits that promote mating within a species, as well as character displacement, which is when two species become more distinct in appearance.

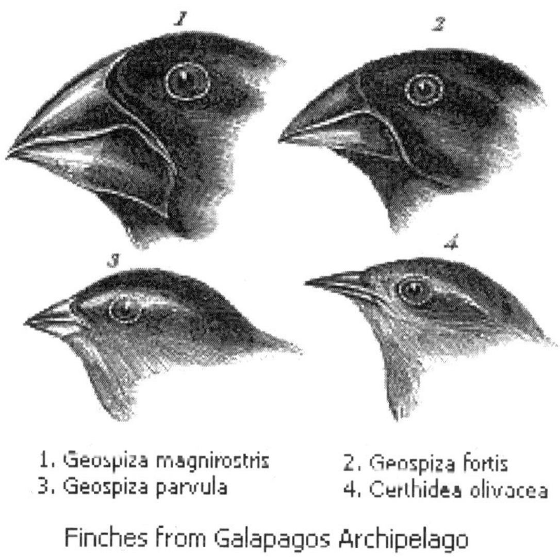

Fig. 7: Geographical isolation of finches on the Galápagos Islands produced over a dozen new species.

Finally, in sympatric speciation species diverge without geographic isolation or changes in habitat. This form is rare since even a small amount of gene flow may remove genetic differences between parts of a population. Generally, sympatric speciation in animals requires the evolution of both genetic differences and non-random mating, to allow reproductive isolation to evolve.

One type of sympatric speciation involves cross-breeding of two related species to produce a new hybrid species. This is not common in animals as animal hybrids are usually sterile. This is because during meiosis the homologous chromosomes from each parent are from different species and cannot successfully pair. However, it is more common in plants because plants often double their number of chromosomes, to form polyploids. This allows the chromosomes from each parental species to form matching pairs during meiosis, since each parent's chromosomes are represented by a pair already. An example of such a speciation event is when the plant species Arabidopsis thaliana and Arabidopsis arenosa cross-bred to give the new species Arabidopsis suecica. This happened about 20,000 years ago, and the speciation process has been repeated in the laboratory, which allows the study of the genetic mechanisms involved in this process. Indeed, chromosome doubling within a species may be a common cause of reproductive isolation, as half the doubled chromosomes will be unmatched when breeding with undoubled organisms.

Speciation events are important in the theory of punctuated equilibrium, which accounts for the pattern in the fossil record of short "bursts" of evolution interspersed with relatively long periods of stasis, where species remain relatively unchanged. In this theory, speciation and rapid evolution are linked, with natural selection and genetic drift acting most strongly on organisms undergoing speciation in novel habitats or small populations. As a result, the periods of stasis in the fossil record correspond to the parental population and the organisms undergoing speciation and rapid evolution are found in small populations or geographically restricted habitats and therefore rarely being preserved as fossils.

Extinction

Fig. 8: Tyrannosaurus rex. Non-avian dinosaursdied out in the Cretaceous–Tertiary extinction event at the end of theCretaceous period.

Extinction is the disappearance of an entire species. Extinction is not an unusual event, as species regularly appear through speciation and disappear through extinction. Nearly all animal and plant species that have lived on Earth are now extinct, and extinction appears to be the ultimate fate of all species.

These extinctions have happened continuously throughout the history of life, although the rate of extinction spikes in occasional mass extinction events. The Cretaceous–Tertiary extinction event, during which the non-avian dinosaurs went extinct, is the most well-known, but the earlier Permian–Triassic extinction event was even more severe, with approximately 96% of species driven to extinction. The Holocene extinction event is an ongoing mass extinction associated with humanity's expansion across the globe over the past few thousand years.

Present-day extinction rates are 100–1000 times greater than the background rate and up to 30% of species may be extinct by the mid 21st century. Human activities are now the primary cause of the ongoing extinction event; global warming may further accelerate it in the future.

The role of extinction in evolution is not very well understood and may depend on which type of extinction is considered. The causes of the continuous "low-level" extinction events, which form the majority of extinctions, may be the result of competition between species for limited resources (competitive exclusion). If one species can out-compete another, this could produce species selection, with the fitter species surviving and the other species being driven to extinction. The intermittent mass extinctions are also important, but instead of acting as a selective force, they drastically reduce diversity in a nonspecific manner and promote bursts of rapid evolution and speciation in survivors.

Evolutionary History of Life

Highly energetic chemistry is believed to have produced a self-replicating molecule around 4 billion years ago and half a billion years later the last common ancestor of all life existed. The currentscientific consensus is that the complex biochemistry that makes up life came from simpler chemical reactions. The beginning of life may have included self-replicating molecules such asRNA, and the assembly of simple cells.

Common Descent

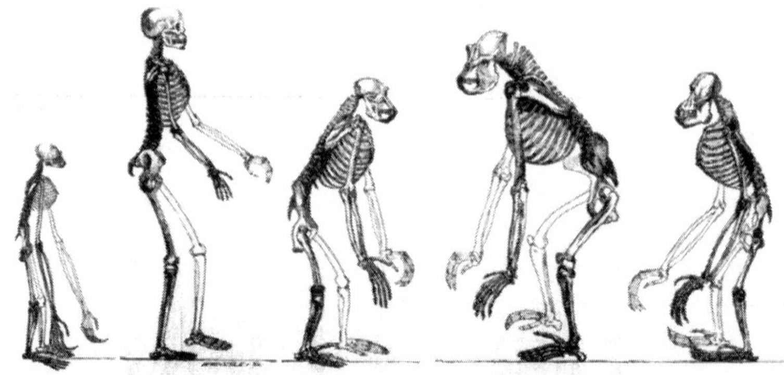

Fig. 9: The hominoids are descendants of a common ancestor.

All organisms on Earth are descended from a common ancestor or ancestral gene pool. Current species are a stage in the process of evolution, with their diversity the product of a long series of speciation and extinction events. The common descent of organisms was first deduced from four simple facts about organisms: First, they have geographic distributions that cannot be explained by local adaptation. Second, the diversity of life is not a set of completely unique organisms, but organisms that share morphological similarities.

Third, vestigial traits with no clear purpose resemble functional ancestral traits and finally, that organisms can be classified using these similarities into a hierarchy of nested groups – similar to a family tree. However, modern research has suggested that, due to horizontal gene transfer, this "tree of life" may be more complicated than a simple branching tree since some genes have spread independently between distantly related species.

Past species have also left records of their evolutionary history. Fossils, along with the comparative anatomy of present-day organisms, constitute the morphological, or anatomical, record. By comparing the anatomies of both modern and extinct species, paleontologists can infer the lineages of those species. However, this approach is most successful for organisms that had hard body parts, such as shells, bones or teeth. Further, as prokaryotes such as bacteria and archaea share a limited set of common morphologies, their fossils do not provide information on their ancestry.

More recently, evidence for common descent has come from the study of biochemical similarities between organisms. For example, all living cells use the same basic set of nucleotides and amino acids. The development of molecular genetics has revealed the record of evolution left in organisms' genomes: dating when species diverged through the molecular clock produced by mutations.For example, these DNA sequence comparisons have revealed that humans and chimpanzees share 96% of their genomes and analyzing the few areas where they differ helps shed light on when the common ancestor of these species existed.

Evolution of Life

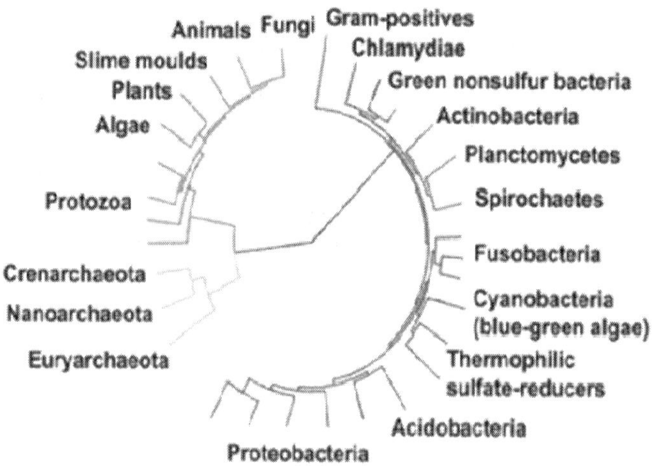

Fig. 10:

Evolutionary tree showing the divergence of modern species from their common ancestor in the centre. The three domains are coloured, withbacteria blue, archaea green and eukaryotes red.

Prokaryotes inhabited the Earth from approximately 3–4 billion years ago. No obvious changes in morphology or cellular organisation occurred in these organisms over the next few billion years.

The eukaryotic cells emerged between 1.6 – 2.7 billion years ago. The next major change in cell structure came when bacteria were engulfed by eukaryotic cells, in a cooperative association called endosymbiosis. The engulfed bacteria and the host cell then underwent co-evolution, with the bacteria evolving into either mitochondria or hydrogenosomes. Another engulfment of cyanobacterial-like organisms led to the formation of chloroplasts in algae and plants.

The history of life was that of the unicellular eukaryotes, prokaryotes and archaea until about 610 million years ago when multicellular organisms began to appear in the oceans in the Ediacaran period. The evolution of multicellularityoccurred in multiple independent events, in organisms as diverse as sponges, brown algae, cyanobacteria, slime moulds andmyxobacteria.

Soon after the emergence of these first multicellular organisms, a remarkable amount of biological diversity appeared over approximately 10 million years, in an event called the Cambrian explosion. Here, the majority of types of modern animals appeared in the fossil record, as well as unique lineages that subsequently became extinct. Various triggers for the Cambrian explosion have been proposed, including the accumulation of oxygen in the atmosphere from photosynthesis.

About 500 million years ago, plants and fungi colonised the land and were soon followed by arthropods and other animals.Insects were particularly successful and even today make up the majority of animal species.

Amphibians first appeared around 364 million years ago, followed by early amniotes, then birds around 155 million years ago (both from "reptile"-like lineages), mammals around 129 million years ago, homininae around 10 million years ago and modern humans around 0.25 million years ago. However, despite the evolution of these large animals, smaller organisms similar to the types that evolved early in this process continue to be highly successful and dominate the Earth, with the majority of both biomass and species being prokaryotes.

Applications

Artificial selection is the intentional selection of traits in a population of organisms. This has been used for thousands of years in the domestication of plants and animals. More recently, such selection has become a vital part of genetic engineering, with selectable markers such as antibiotic resistance genes being used to manipulate DNA. In repeated rounds of mutation and selection proteins with valuable properties have evolved, for example modified enzymes and new antibodies, in a process called directed evolution.

Understanding the changes that have occurred during organism's evolution can reveal the genes needed to construct parts of the body, genes which may be involved in human genetic disorders. For example, the mexican tetra is an albino cavefish that lost its eyesight during evolution. Breeding together different populations of this blind fish produced some offspring with functional eyes, since different mutations had occurred in the isolated populations that had evolved in different caves. This helped identify genes required for vision and pigmentation.

In computer science, simulations of evolution using evolutionary algorithms and artificial life started in the 1960s and was extended with simulation of artificial selection. Artificial evolution became a widely recognised optimisation method as a result of the work of Ingo Rechenberg in the 1960s. He used evolution strategies to solve complex engineering problems.

Genetic algorithms in particular became popular through the writing of John Holland. Practical applications also include automatic evolution of computer programs. Evolutionary algorithms are now used to solve multi-dimensional problems more efficiently than software produced by human designers and also to optimise the design of systems.

Social and Cultural Responses

Fig. 11: As evolution became widely accepted in the 1870s,caricatures of Charles Darwinwith an ape or monkey body symbolised evolution.

In the 19th century, particularly after the publication of On the Origin of Species in 1859, the idea that life had evolved was an active source of academic debate centred on the philosophical, social and religious implications of evolution.

Nowadays, the modern evolutionary synthesis is accepted by a vast majority of scientists. However, evolution remains a contentious concept for some theists.

While various religions and denominations have reconciled their beliefs with evolution through concepts such as theistic evolution, there are creationists who believe that evolution is contradicted by the creation myths found in their religions and who raise various objections to evolution.

As had been demonstrated by responses to the publication of Vestiges of the Natural History of Creation in 1844, the most controversial aspect of evolutionary biology is the implication of human evolution that humans share common ancestry with apes and that the mental and moral faculties of humanity have the same types of natural causes as other inherited traits in animals.

In some countries, notably the United States, these tensions between science and religion have fuelled the current creation-evolution controversy, a religious conflict focusing on politics and public education. While other scientific fields such as cosmology and Earth science also conflict with literal interpretations of many religious texts, evolutionary biology experiences significantly more opposition from religious literalists.

The teaching of evolution in American secondary school biology classes was uncommon in most of the first half of the 20th century.

The Scopes Trial decision of 1925 caused the subject to become very rare in American secondary biology textbooks for a generation, but it was gradually re-introduced about a generation later and legally protected with the 1968 Epperson v. Arkansas decision.

Since then, the competing religious belief of creationism was legally disallowed in secondary school curricula in various decisions in the 1970s and 1980s, but it returned in the form of intelligent design, to be excluded once again in the 2005 Kitzmiller v. Dover Area School District case.

Molecular Evolution

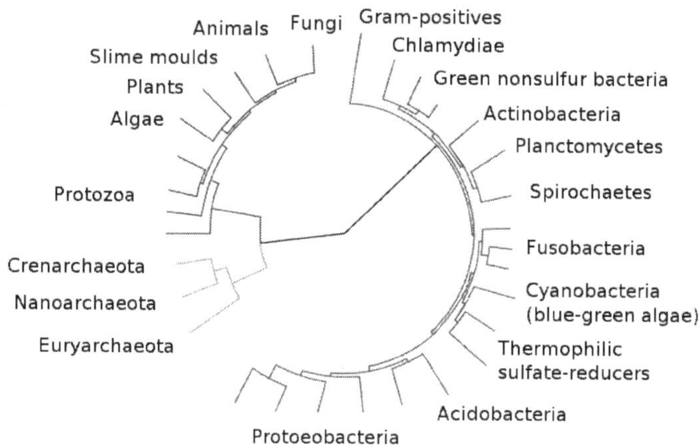

Fig. 12:

Molecular evolution is in part a process of evolution at the scale of DNA, RNA, and proteins. Molecular evolution emerged as a scientific field in the 1960s as researchers from molecular biology, evolutionary biology and population genetics sought to understand recent discoveries on the structure and function of nucleic acids and protein. Some of the key topics that spurred development of the field have been the evolution of enzyme function, the use of nucleic acid divergence as a "molecular clock" to study species divergence, and the origin of noncoding DNA.

Recent advances in genomics, including whole-genome sequencing, high-throughput protein characterization, and bioinformatics have led to a dramatic increase in studies on the topic. In the 2000s, some of the active topics have been the role of gene duplication in the emergence of novel gene function, the extent of adaptive molecular evolution versus neutral processes of mutation and drift, and the identification of molecular changes responsible for various human characteristics especially those pertaining to infection, disease, and cognition.

Principles of Molecular Evolution

Mutations

Mutations are permanent, transmissible changes to the genetic material (usually DNA or RNA) of a cell. Mutations can be caused by copying errors in the genetic material during cell division and by exposure to radiation, chemicals, or viruses, or can occur deliberately under cellular control during the processes such as meiosis or hypermutation.

Mutations are considered the driving force ofevolution, where less favorable (or deleterious) mutations are removed from the gene pool by natural selection, while more favorable (or beneficial) ones tend to accumulate. Neutral mutations do not affect the organism's chances of survival in its natural environment and can accumulate over time, which might result in what is known as punctuated equilibrium; the modern interpretation of classic evolutionary theory.

Causes of Change in Allele Frequency

There are four known processes that affect the survival of a characteristic; or, more specifically, the frequency of an allele (variant of a gene):

- Genetic drift describes changes in gene frequency that cannot be ascribed to selective pressures, but are due instead to events that are unrelated to inherited traits. This is especially important in small mating populations, which simply cannot have enough offspring to maintain the same gene distribution as the parental generation.
- Gene flow or Migration: or gene admixture is the only one of the agents that makes populations closer genetically while building larger gene pools.
- Selection, in particular natural selection produced by differential mortality and fertility. Differential mortality is the survival rate of individuals before their reproductive age. If they survive, they are then selected further by differential fertility – that is, their total genetic contribution to the next generation. In this way, the alleles that these surviving individuals contribute to the gene pool will increase the frequency of those alleles. Sexual selection, the attraction between mates that results from two genes, one for a feature and the other determining a preference for that feature, is also very important.
- Recurrent mutation can increase the frequency of a mutant allele.

Molecular Study of Phylogeny

Molecular systematics is a product of the traditional field of systematics and molecular genetics. It is the process of using data on the molecular constitution of biological organisms' DNA, RNA, or both, in order to resolve questions in systematics, i.e. about their correct scientific classification or taxonomy from the point of view of evolutionary biology.

Molecular systematics has been made possible by the availability of techniques for DNA sequencing, which allow the determination of the exact sequence of nucleotides or bases in either DNA or RNA. At present it is still a long and expensive process to sequence the entire genome of an organism, and this has been done for only a few species. However, it is quite feasible to determine the sequence of a defined area of a particular chromosome. Typical molecular systematic analyses require the sequencing of around 1000 base pairs.

The Driving Forces of Evolution

Depending on the relative importance assigned to the various forces of evolution, three perspectives provide evolutionary explanations for molecular evolution.

While recognizing the importance of random drift for silent mutations, selectionists hypotheses argue that balancing and positive selection are the driving forces of molecular evolution. Those hypotheses are often based on the broader view called panselectionism, the idea that selection is the only force strong enough to explain evolution, relaying random drift and mutations to minor roles.

Neutralists hypotheses emphasize the importance of mutation, purifying selection and random genetic drift. The introduction of the neutral theory by Kimura, quickly followed by King and Jukes' own findings, led to a fierce debate about the relevance of neodarwinism at the molecular level. The Neutral theory of molecular evolution states that most mutations are deleterious and quickly removed by natural selection, but of the remaining ones, the vast majority are neutral with respect to fitness while the amount of advantageous mutations is vanishingly small. The fate of neutral mutations are governed by genetic drift, and contribute to both nucleotide polymorphism and fixed differences between species.

Mutationists hypotheses emphasize random drift and biases in mutation patterns. Sueoka was the first to propose a modern mutationist view. He proposed that the variation in GC content was not the result of positive selection, but a consequence of the GC mutational pressure.

History of the Science

The history of molecular evolution starts in the early 20th century with "comparative biochemistry", but the field of molecular evolution came into its own in the 1960s and 1970s, following the rise ofmolecular biology. The advent of protein sequencing allowed molecular biologists to create phylogenies based on sequence comparison, and to use the differences between homologous sequences as a molecular clock to estimate the time since the last common ancestor. In the late 1960s, the neutral theory of molecular evolution provided a theoretical basis for the molecular clock, though both the clock and the neutral theory were controversial, since most evolutionary biologists held strongly to panselectionism, with natural selection as the only important cause of evolutionary change. After the 1970s, nucleic acid sequencing allowed molecular evolution to reach beyond proteins to highly conserved ribosomal RNA sequences, the foundation of a reconceptualization of the early history of life.

The theoretical frameworks for molecular systematics were laid in the 1960s in the works of Emile Zuckerkandl, Emanuel Margoliash, Linus Pauling and Walter M. Fitch. Applications of molecular systematics were pioneered by Charles G. Sibley (birds), Herbert C. Dessauer (herpetology), and Morris Goodman (primates), followed by Allan C. Wilson, Robert K. Selander, and John C. Avise (who studied various groups). Work with protein electrophoresis began around 1956. Although the results were not quantitative and did not initially improve on morphological classification, they provided tantalizing hints that long-held notions of the classifications of birds, for example, needed substantial revision. In the period of 1974–1986, DNA-DNA hybridization was the dominant technique.

Chapter 6

Genome Evolution

Genomic evolution is a set of phenomena involved in the changing of the structure of a genome through evolution.

The study of genome evolution involves multiple fields such as structural analysis of the genome, the study of genomic parasites, gene and ancient genome duplications, polyploidy, and comparative genomics. Evolutionary biologists are interested in five specific questions in regards to evolution of the genome, these are:

1. How did the genome evolve into its current size?
2. What is the content within the genome, is it mostly junk or not?
3. What is the distribution of genes within a genome?
4. What is the composition of the nucleotides within the genome?
5. How does translation of the genetic code evolve?

Genome Size

Genome size is all the DNA that makes the genome. A genome can consist of genetic regions and noncoding regions. Genetic regions are those that encode proteins while noncoding regions refer to promoters and junk DNA. The C-value is another term for the genome size. Within a species the C-value does not show much variation, but there is a significant difference in the C-value between species.

Prokaryotic Genome

Prokaryotes are unicellular organisms that do not have membrane-bound organelles and lack a structurally distinct nucleus. Research on prokaryotic genomes shows that there is a significant positive correlation between the C-value of prokaryotes and the amount of genes that compose the genome. This indicates that gene size is the main factor influencing the size of the genome.

Eukaryotic Genome

In eukaryotic organisms, there is a paradox observed, namely that the number of genes that make up the genome does not correlate with genome size. In other words, the genome size is much larger than would be expected given the total number of protein coding genes.

Related Fields

An important area within the study of molecular evolution is the use of molecular data to determine the correct biological classification of organisms. This is called molecular systematics or molecular phylogenetics.

Tools and concepts developed in the study of molecular evolution are now commonly used for comparative genomics and molecular genetics, while the influx of new data from these fields has been spurring advancement in molecular evolution.

Key Researchers in Molecular Evolution

Some researchers who have made key contributions to the development of the field:

- Motoo Kimura — Neutral theory
- Masatoshi Nei — Adaptive evolution
- Walter M. Fitch — Phylogenetic reconstruction
- Walter Gilbert — RNA world
- Joe Felsenstein — Phylogenetic methods
- Susumu Ohno — Gene duplication
- John H. Gillespie — Mathematics of adaptation
- Dan Graur - Neutral models of molecular evolution
- Wen-Hsiung Li - Neutral models of molecular evolution

Journals and Societies

Journals dedicated to molecular evolution include Molecular Biology and Evolution, Journal of Molecular Evolution, and Molecular Phylogenetics and Evolution. Research in molecular evolution is also published in journals of genetics, molecular biology, genomics, systematics, or evolutionary biology. The Society for Molecular Biology and Evolution publishes the journal "Molecular Biology and Evolution" and holds an annual international meeting.

Population Genetics

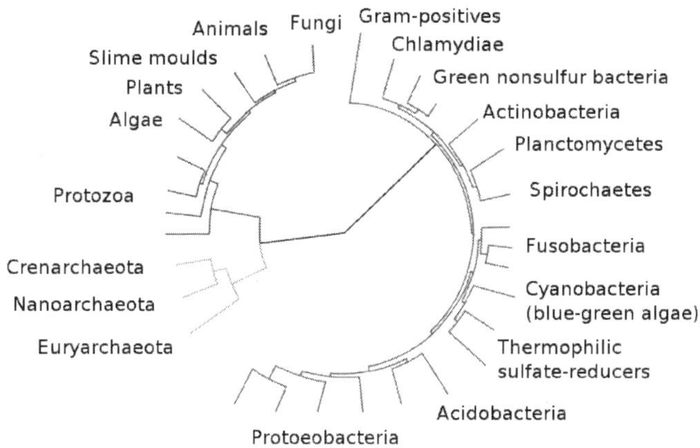

Fig. 1:

Key Topic

History of Theory

Natural History

Processes & Outcomes

Social Implications

Fields & Applications

Population genetics is the study of allele frequency distribution and change under the influence of the four main evolutionary processes: natural selection,genetic drift, mutation and gene flow. It also takes into account the factors of recombination, population subdivision and population structure. It attempts to explain such phenomena as adaptation and speciation.

Population genetics was a vital ingredient in the emergence of the modern evolutionary synthesis. Its primary founders were Sewall Wright, J. B. S. Haldaneand R. A. Fisher, who also laid the foundations for the related discipline of quantitative genetics.

Fundamentals

Population genetics is the study of the frequency and interaction of alleles and genes in populations. A population is a set of organisms in which any pair of members can breed together. This implies that all members belong to the same species and live near each other.

For example, all of the moths of the same species living in an isolated forest are a population. A gene in this population may have several alternate forms, which account for variations between thephenotypes of the organisms. An example might be a gene for coloration in moths that has two alleles: black and white. A gene pool is the complete set of alleles for a gene in a single population; theallele frequency for an allele is the fraction of the genes in the pool that is composed of that allele (for example, what fraction of moth coloration genes are the black allele). Evolution occurs when there are changes in the frequencies of alleles within a population; for example, the allele for black color in a population of moths becoming more common.

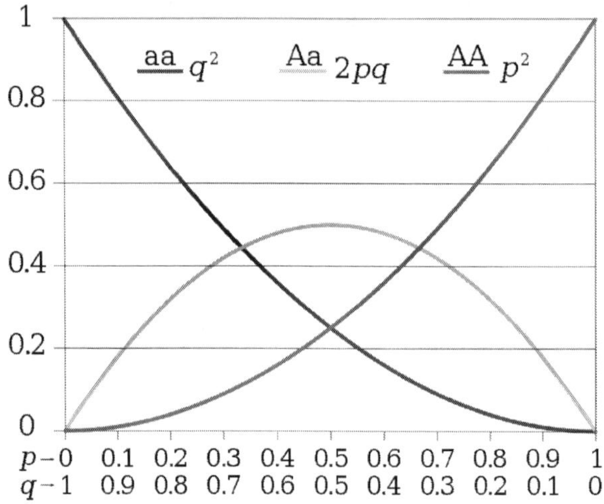

Fig. 2:

Hardy–Weinberg principle for two alleles: the horizontal axis shows the two allele frequencies p and q and the vertical axis shows the genotype frequencies. Each graph shows one of the three possible genotypes.

Hardy–Weinberg Principle

Natural selection will only cause evolution if there is enough genetic variation in a population. Before the discovery of Mendelian genetics, one common hypothesis was blending inheritance. But with blending inheritance, genetic variance would be rapidly lost, making evolution by natural selection implausible. The Hardy-Weinberg principle provides the solution to how variation is maintained in a population withMendelian inheritance. According to this principle, the frequencies of alleles (variations in a gene) will remain constant in the absence of selection, mutation, migration and genetic drift. The Hardy-Weinberg "equilibrium" refers to this stability of allele frequencies over time.

A second component of the Hardy-Weinberg principle concerns the effects of a single generation of random mating. In this case, the genotype frequencies can be predicted from the allele frequencies. For example, in the simplest case of a single locus with two alleles: thedominant allele is denoted A and the recessive a and their frequencies are denoted by p and q; freq(A) = p; freq(a) = q; p + q = 1. If the genotype frequencies are in Hardy-Weinberg proportions resulting from random mating, then we will have freq(AA) = p for the AAhomozygotes in the population, freq(aa) = q for the aa homozygotes, and freq(Aa) = 2pq for the heterozygotes.

Genetic Genealogy

Concepts

" Population genetics

" Haplogroup/ Haplotype

" Most recent common ancestor

" Human mitochondrial DNA haplogroups

" Human Y-chromosome DNA haplogroups

" Genomics

Other

" Y-chromosome haplogroups by populations

" Genealogical DNA test

" Personal genomics

" Genographic Project

This box: view · talk · edit

Scope and Theoretical Considerations

The mathematics of population genetics were originally developed as part of the modern evolutionary synthesis. According to Beatty (1986), it defines the core of the modern synthesis.

According to Lewontin (1974), the theoretical task for population genetics is a process in two spaces: a "genotypic space" and a "phenotypic space". The challenge of a complete theory of population genetics is to provide a set of laws that predictably map a population of genotypes (G_1) to aphenotype space (P_1), where selection takes place, and another set of laws that map the resulting population (P_2) back to genotype space (G_2) where Mendeliangenetics can predict the next generation of genotypes, thus completing the cycle. Even leaving aside for the moment the non-Mendelian aspects of molecular genetics, this is clearly a gargantuan task. Visualizing this transformation schematically:

$$G_1 \xrightarrow{T_1} P_1 \xrightarrow{T_2} P_2 \xrightarrow{T_3} G_2 \xrightarrow{T_4} G'_1 \rightarrow \cdots$$

(adapted from Lewontin 1974, p. 12). XD

T_1 represents the genetic and epigenetic laws, the aspects of functional biology, or development, that transform a genotype into phenotype. We will refer to this as the "genotype-phenotype map". T_2 is the transformation due to natural selection, T_3 are epigenetic relations that predict genotypes based on the selected phenotypes and finally T_4 the rules of Mendelian genetics.

In practice, there are two bodies of evolutionary theory that exist in parallel, traditional population genetics operating in the genotype space and the biometric theory used in plant and animal breeding, operating in phenotype space. The missing part is the mapping between the genotype and phenotype space. This leads to a "sleight of hand" (as Lewontin terms it) whereby variables in the equations of one domain, are considered parameters or constants, where, in a full-treatment they would be transformed themselves by the evolutionary process and are in reality functions of the state variables in the other domain. The "sleight of hand" is assuming that we know this mapping. Proceeding as if we do understand it is enough to analyze many cases of interest. For example, if the phenotype is almost one-to-one with genotype (sickle-cell disease) or the time-scale is sufficiently short, the "constants" can be treated as such; however, there are many situations where it is inaccurate.

The Four Processes

Natural Selection

Natural selection is the fact that some traits make it more likely for an organism to survive and reproduce. Population genetics describes natural selection by defining fitness as a propensity or probability of survival and reproduction in a particular environment. The fitness is normally given by the symbol $w=1+s$ where s is the selection coefficient. Natural selection acts on phenotypes, or the observable characteristics of organisms, but the genetically heritable basis of any phenotype which gives a reproductive advantage will become more common in a population. In this way, natural selection converts differences in fitness into changes in allele frequency in a population over successive generations.

Before the advent of population genetics, many biologists doubted that small difference in fitness were sufficient to make a large difference to evolution. Population geneticists addressed this concern in part by comparing selection to genetic drift. Selection can overcome genetic drift when s is greater than 1 divided by the effective population size. When this criterion is met, the probability that a new advantageous mutant becomes fixed is approximately equal to s. The time until fixation of such an allele depends little on genetic drift, and is approximately proportional to $\log(sN)/s$.

Genetic Drift

Genetic drift is a change in allele frequencies caused by random sampling. That is, the alleles in the offspring are a random sample of those in the parents. Genetic drift may cause gene variants to disappear completely, and thereby reduce genetic variability. In contrast to natural selection, which makes gene variants more common or less common depending on their reproductive success, the changes due to genetic drift are not driven by environmental or adaptive pressures, and may be beneficial, neutral, or detrimental to reproductive success.

The effect of genetic drift is larger for alleles present in a smaller number of copies, and smaller when an allele is present in many copies. Vigorous debates wage among scientists over the relative importance of genetic drift compared with natural selection. Ronald Fisher held the view that genetic drift plays at the most a minor role in evolution, and this remained the dominant view for several decades. In 1968 Motoo Kimura rekindled the debate with his neutral theory of molecular evolution which claims that most of the changes in the genetic material are caused by neutral mutations and genetic drift. The role of genetic drift by means of sampling error in evolution has been criticized by John H Gillespie and Will Provine, who argue that selection on linked sites is a more important stochastic force.

The population genetics of genetic drift are described using either branching processes or a diffusion equation describing changes in allele frequency. These approaches are usually applied to the Wright-Fisher and Moran models of population genetics. Assuming genetic drift is the only evolutionary force acting on an allele, after t generations in many replicated populations, starting with allele frequencies of p and q, the variance in allele frequency across those populations is

$$V_t \approx pq\left(1-\exp\left\{-\frac{t}{2N_e}\right\}\right)$$

Mutation

Mutation is the ultimate source of genetic variation in the form of new alleles. Mutation can result in several different types of change in DNA sequences; these can either have no effect, alter the product of a gene, or prevent the gene from functioning. Studies in the fly Drosophila melanogaster suggest that if a mutation changes a protein produced by a gene, this will probably be harmful, with about 70 percent of these mutations having damaging effects, and the remainder being either neutral or weakly beneficial.

Mutations can involve large sections of DNA becoming duplicated, usually through genetic recombination. These duplications are a major source of raw material for evolving new genes, with tens to hundreds of genes duplicated in animal genomes every million years. Most genes belong to larger families of genes of shared ancestry. Novel genes are produced by several methods, commonly through the duplication and mutation of an ancestral gene, or by recombining parts of different genes to form new combinations with new functions.

Here, domains act as modules, each with a particular and independent function, that can be mixed together to produce genes encoding new proteins with novel properties. For example, the human eye uses four genes to make structures that sense light: three for color vision and one for night vision; all four arose from a single ancestral gene.

Another advantage of duplicating a gene (or even an entire genome) is that this increasesredundancy; this allows one gene in the pair to acquire a new function while the other copy performs the original function. Other types of mutation occasionally create new genes from previously noncoding DNA.

In addition to being a major source of variation, mutation may also function as a mechanism of evolution when there are different probabilities at the molecular level for different mutations to occur, a process known as mutation bias.

If two genotypes, for example one with the nucleotide G and another with the nucleotide A in the same position, have the same fitness, but mutation from G to A happens more often than mutation from A to G, then genotypes with A will tend to evolve. Different insertion vs. deletion mutation biases in different taxa can lead to the evolution of different genome sizes. Developmental or mutational biases have also been observed in morphological evolution. For example, according to the phenotype-first theory of evolution, mutations can eventually cause the genetic assimilation of traits that were previously induced by the environment.

Mutation bias effects are superimposed on other processes. If selection would favor either one out of two mutations, but there is no extra advantage to having both, then the mutation that occurs the most frequently is the one that is most likely to become fixed in a population. Mutations leading to the loss of function of a gene are much more common than mutations that produce a new, fully functional gene. Most loss of function mutations are selected against. But when selection is weak, mutation bias towards loss of function can affect evolution. For example, pigments are no longer useful when animals live in the darkness of caves, and tend to be lost.

This kind of loss of function can occur because of mutation bias, and/or because the function had a cost, and once the benefit of the function disappeared, natural selection leads to the loss. Loss of sporulation ability in a bacterium during laboratory evolution appears to have been caused by mutation bias, rather than natural selection against the cost of maintaining sporulation ability. When there is no selection for loss of function, the speed at which loss evolves depends more on the mutation rate than it does on theeffective population size, indicating that it is driven more by mutation bias than by genetic drift.

Evolution of Mutation Rate

Due to the damaging effects that mutations can have on cells, organisms have evolved mechanisms such as DNA repair to remove mutations. Therefore, the optimal mutation rate for a species is a trade-off between costs of a high mutation rate, such as deleterious mutations, and the metabolic costs of maintaining systems to reduce the mutation rate, such as DNA repair enzymes.

Viruses that use RNA as their genetic material have rapid mutation rates, which can be an advantage since these viruses will evolve constantly and rapidly, and thus evade the defensive responses of e.g. the human immune system.

Gene Flow & Transfer

Gene flow is the exchange of genes between populations, which are usually of the same species. Examples of gene flow within a species include the migration and then breeding of organisms, or the exchange of pollen. Gene transfer between species includes the formation of hybrid organisms and horizontal gene transfer.

Gene Flow

Migration into or out of a population can change allele frequencies, as well as introducing genetic variation into a population. Immigration may add new genetic material to the established gene pool of a population. Conversely, emigration may remove genetic material. As barriers to reproduction between two diverging populations are required for the populations to become new species, gene flow may slow this process by spreading genetic differences between the populations. Gene flow is hindered by mountain ranges, oceans and deserts or even man-made structures such as the Great Wall of China, which has hindered the flow of plant genes.

Depending on how far two species have diverged since their most recent common ancestor, it may still be possible for them to produce offspring, as with horses and donkeys mating to producemules. Such hybrids are generally infertile, due to the two different sets of chromosomes being unable to pair up during meiosis.

In this case, closely related species may regularly interbreed, but hybrids will be selected against and the species will remain distinct. However, viable hybrids are occasionally formed and these new species can either have properties intermediate between their parent species, or possess a totally new phenotype. The importance of hybridization in creating new species of animals is unclear, although cases have been seen in many types of animals, with thegray tree frog being a particularly well-studied example.

Hybridization is, however, an important means of speciation in plants, since polyploidy (having more than two copies of each chromosome) is tolerated in plants more readily than in animals.Polyploidy is important in hybrids as it allows reproduction, with the two different sets of chromosomes each being able to pair with an identical partner during meiosis. Polyploids also have more genetic diversity, which allows them to avoid inbreeding depression in small populations.

Horizontal Gene Transfer

Horizontal gene transfer is the transfer of genetic material from one organism to another organism that is not its offspring; this is most common among bacteria. In medicine, this contributes to the spread of antibiotic resistance, as when one bacteria acquires resistance genes it can rapidly transfer them to other species. Horizontal transfer of genes from bacteria to eukaryotes such as the yeast Saccharomyces cerevisiae and the adzuki bean beetle Callosobruchus chinensis may also have occurred.

An example of larger-scale transfers are the eukaryotic bdelloid rotifers, which appear to have received a range of genes from bacteria, fungi, and plants. Viruses can also carry DNA between organisms, allowing transfer of genes even across biological domains. Large-scale gene transfer has also occurred between the ancestors of eukaryotic cells and prokaryotes, during the acquisition of chloroplasts and mitochondria.

Recombination

Basic models of population genetics consider only one gene locus at a time. Because of epistasis, the phenotypic effect of an allele at one locus may depend on which alleles are present at many other loci. If all genes are in linkage equilibrium, the effect of an allele at one locus can be averaged across the gene pool at other loci. In reality, one allele is frequently found in linkage disequilibriumwith genes at other loci, especially with genes located nearby on the same chromosome.

Recombination breaks up this linkage disequilibrium too slowly to avoid genetic hitchhiking, where an allele at one locus rises to high frequency because it is linked to an allele under selection at a nearby locus. This is a problem for population genetic models that treat one gene locus at a time. It can, however, be exploited as a method for detecting the action of natural selection via selective sweeps. In the extreme case of primarily asexual populations, different population genetic equations can be derived and solved, and these behave quite differently to the sexual case.

Genetic Structure

Because of physical barriers to migration, along with limited tendency for individuals to move or spread (vagility), and tendency to remain or come back to natal place (philopatry), natural populations rarely all interbreed as convenient in theoretical random models (panmixy) (Buston et al., 2007). There is usually a geographic range within which individuals are more closely related to one another than those randomly selected from the general population. This is described as the extent to which a population is genetically structured (Repaci et al., 2007). Genetic structuring can be caused by migration due to historical climate change, species range expansion or current availability of habitat.

Microbial Population Genetics

Microbial population genetics is a rapidly advancing field of investigation with relevance to many other theoretical and applied areas of scientific investigations. The population genetics of microorganismslays the foundations for tracking the origin and evolution of antibiotic resistance and deadly infectious pathogens. Population genetics of microorganisms is also an essential factor for devising strategies for the conservation and better utilization of beneficial microbes (Xu, 2010).

History

Fig. 3: Biston betularia f. typica is the white-bodied form of the peppered moth.

Fig. 4: Biston betularia f. carbonaria is the black-bodied form of the peppered moth.

Population Genetics

Population genetics was developed as a reconciliation of the Mendelian and biometrician models. A key step was the work of the British biologist and statisticianR.A. Fisher.

In a series of papers starting in 1918 and culminating in his 1930 book The Genetical Theory of Natural Selection, Fisher showed that the continuous variation measured by the biometricians could be produced by the combined action of many discrete genes, and that natural selection could changegene frequencies in a population, resulting in evolution (though lacking the knowledge of what an actual gene was at this time, it should be said in this sense he understood phenotypic trait frequency, rather than specifically identifiable gene frequency).

In a series of papers beginning in 1924, another British geneticist,J.B.S. Haldane, applied statistical analysis to real-world examples of natural selection, such as the evolution of industrial melanism in peppered moths, and showed that natural selection worked at an even faster rate than Fisher assumed.

The American biologist Sewall Wright, who had a background in animal breeding experiments, focused on combinations of interacting genes, and the effects of inbreeding on small, relatively isolated populations that exhibited genetic drift. In 1932, Wright introduced the concept of an adaptive landscape and argued that genetic drift and inbreeding could drive a small, isolated sub-population away from an adaptive peak, allowing natural selection to drive it towards different adaptive peaks. Fisher and Wright had some fundamental disagreements and a controversy about the relative roles of selection and drift continued for much of the century between the Americans and the British. The Frenchman Gustave Malécot was also important early in the development of the discipline.

The work of Fisher, Haldane and Wright founded the discipline of population genetics. This integrated natural selection with Mendelian genetics, which was the critical first step in developing a unified theory of how evolution worked.

John Maynard Smith was Haldane's pupil, whilst W.D. Hamilton was heavily influenced by the writings of Fisher. The American George R. Price worked with both Hamilton and Maynard Smith. American Richard Lewontin and Japanese Motoo Kimura were heavily influenced by Wright.

Modern Evolutionary Synthesis

In the first few decades of the 20th century, most field naturalists continued to believe that Lamarckian and orthogenic mechanisms of evolution provided the best explanation for the complexity they observed in the living world. However, as the field of genetics continued to develop, those views became less tenable. Theodosius Dobzhansky, a postdoctoral worker in T. H. Morgan's lab, had been influenced by the work on genetic diversity by Russian geneticists such as Sergei Chetverikov. He helped to bridge the divide between the foundations of microevolution developed by the population geneticists and the patterns of macroevolution observed by field biologists, with his 1937 book Genetics and the Origin of Species.

Dobzhansky examined the genetic diversity of wild populations and showed that, contrary to the assumptions of the population geneticists, these populations had large amounts of genetic diversity, with marked differences between sub-populations. The book also took the highly mathematical work of the population geneticists and put it into a more accessible form. In Great Britain E.B. Ford, the pioneer of ecological genetics, continued throughout the 1930s and 1940s to demonstrate the power of selection due to ecological factors including the ability to maintain genetic diversity throughgenetic polymorphisms such as human blood types. Ford's work would contribute to a shift in emphasis during the course of the modern synthesis towards natural selection over genetic drift. Nevertheless, ongoing studies of eukaryotic transposable elements, and of their impact on speciation, point again to a major role of nonadaptive processes such as mutation andgenetic drift. Mutation and genetic drift are also viewed as major factors in the evolution of genome complexity

Mendelian Inheritance

Mendelian inheritance (or Mendelian genetics or Mendelism) is a scientific description of how hereditary characteristics are passed from parent organisms to their offspring; it underlies much of genetics. This theoretical framework was initially derived from the work of Gregor Johann Mendel published in 1865 and 1866 which was re-discovered in 1900; it was initially very controversial. When Mendel's theories were integrated with the chromosome theory of inheritance by Thomas Hunt Morgan in 1915, they became the core of classical genetics.

History

The laws of inheritance were derived by Gregor Mendel, a 19th century Austrian Priest/monk conducting hybridization experiments in garden peas (Pisum sativum). Between 1856 and 1863, he cultivated and tested some 29,000 pea plants. From these experiments he deduced two generalizations which later became known as Mendel's Principles of Heredity or Mendelian inheritance. He described these principles in a two part paper, Experiments on Plant Hybridization that he read to the Natural History Society of Brnoon February 8 and March 8, 1865, and which was published in 1866.

Mendel's conclusions were largely ignored. Although they were not completely unknown to biologists of the time, they were not seen as generally applicable, even by Mendel himself, who thought they only applied to certain categories of species or traits. A major block to understanding their significance was the importance attached by 19th century biologists to the apparent blending of inherited traits in the overall appearance of the progeny, now known to be due to multigene interactions, in contrast to the organ-specific binary characters studied by Mendel. In 1900, however, his work was "re-discovered" by three European scientists, Hugo de Vries, Carl Correns, and Erich von Tschermak. The exact nature of the "re-discovery" has been somewhat debated: De Vries published first on the subject, mentioning Mendel in a footnote, while Correns pointed out Mendel's priority after having read De Vries's paper and realizing that he himself did not have priority. De Vries may not have acknowledged truthfully how much of his knowledge of the laws came from his own work, or came only after reading Mendel's paper. Later scholars have accused Von Tschermak of not truly understanding the results at all. Regardless, the "re-discovery" made Mendelism an important but controversial theory. Its most vigorous promoter in Europe was William Bateson, who coined the term "genetics" and "allele" to describe many of its tenets. The model of heredity was highly contested by other biologists because it implied that heredity was discontinuous, in opposition to the apparently continuous variation observable for many traits. Many biologists also dismissed the theory because they were not sure it would apply to all species. However later work by biologists and statisticians such as R.A. Fisher showed that if multiple Mendelian factors were involved in the expression of an individual trait, they could produce the diverse results observed.

Thomas Hunt Morgan and his assistants later integrated the theoretical model of Mendel with the chromosome theory of inheritance, in which the chromosomes of cells were thought to hold the actual hereditary material, and create what is now known as classical genetics, which was extremely successful and cemented Mendel's place in history. Mendel's findings allowed other scientists to predict the expression of traits on the basis of mathematical probabilities. A large contribution to Mendel's success can be traced to his decision to start his crosses only with plants he demonstrated were true-breeding. He also only measured absolute (binary) characteristics, such as color, shape, and position of the offspring, rather than quantitative characteristics. He expressed his results numerically and subjected them to statistical analysis. His method of data analysis and his large sample size gave credibility to his data. He also had the foresight to follow several successive generations (f2, f3) of his pea plants and record their variations. Finally, he performed "test crosses" (back-crossing descendants of the initial hybridization to the initial true-breeding lines) to reveal the presence and proportion of recessive characters. Without his hard work and careful attention to procedure and detail, Mendel's work could not have had the impact it made on the world of genetics.

Mendel's Laws

Mendel discovered that when crossing white flower and purple flower plants, the result is not a blend. Rather than being a mix of the two, the offspring was purple flowered. He then conceived the idea of heredity units, which he called "factors", one of which is a recessive characteristic and the other dominant. Mendel said that factors, later called genes, normally occur in pairs in ordinary body cells, yet segregate during the formation of sex cells. Each member of the pair becomes part of the separate sex cell. The dominant gene, such as the purple flower in Mendel's plants, will hide the recessive gene, the white flower. After Mendel self-fertilized the F1 generation and obtained the 3:1 ratio, he correctly theorized that genes can be paired in three different ways for each trait: AA, aa, and Aa. The capital "A" represents the dominant factor and lowercase "a" represents the recessive. (The last combination listed above, Aa, will occur roughly twice as often as each of the other two, as it can be made in two different ways, Aa or aA.)

Mendel stated that each individual has two factors for each trait, one from each parent. The two factors may or may not contain the same information. If the two factors are identical, the individual is called homozygous for the trait. If the two factors have different information, the individual is called heterozygous. The alternative forms of a factor are called alleles. The genotype of an individual is made up of the many alleles it possesses. An individual's physical appearance, or phenotype, is determined by its alleles as well as by its environment. An individual possesses two alleles for each trait; one allele is given by the female parent and the other by the male parent. They are passed on when an individual matures and produces gametes: egg and sperm. When gametes form, the paired alleles separate randomly so that each gamete receives a copy of one of the two alleles. The presence of an allele doesn't promise that the trait will be expressed in the individual that possesses it. In heterozygous individuals the only allele that is expressed is the dominant. The recessive allele is present but its expression is hidden.

Mendel summarized his findings in two laws; the Law of Segregation and the Law of Independent Assortment.

Law of Segregation (The "First Law")

The Law of Segregation states that every individual possesses a pair of alleles for any particular trait and that each parent passes a randomly selected copy (allele) of only one of these to its offspring. The offspring then receives its own pair of alleles for that trait. Whichever of the two alleles in the offspring is dominant determines how the offspring expresses that trait (e.g. the color of a plant, the color of an animal's fur, the color of a person's eyes).

More precisely the law states that when any individual produces gametes, the copies of a gene separate so that each gamete receives only one copy (allele). A gamete will receive one allele or the other. The direct proof of this was later found following the observation of meiosis by two independent scientists, the German botanist, Oscar Hertwig in 1876, and the Belgian zoologist, Edouard Van Beneden in 1883. In meiosis, the paternal and maternal chromosomes get separated and the alleles with the traits of a character are segregated into two different gametes.

The two coexisting alleles of an individual for each trait segregate (separate) during gamete formation so that each gamete gets only one of the two alleles. Alleles again unite at random fertilization of gametes.

Law of Independent Assortment (The "Second Law")

The Law of Independent Assortment, also known as "Inheritance Law" states that separate genes for separate traits are passed independently of one another from parents to offspring. That is, the biological selection of a particular gene in the gene pair for one trait to be passed to the offspring has nothing to do with the selection of the gene for any other trait. More precisely the law states that alleles of different genes assort independently of one another during gamete formation. While Mendel's experiments with mixing one trait always resulted in a 3:1 ratio between dominant and recessive phenotypes, his experiments with mixing two traits (dihybrid cross) showed 9:3:3:1 ratios. But the 9:3:3:1 table shows that each of the two genes are independently inherited with a 3:1 phenotypic ratio. Mendel concluded that different traits are inherited independently of each other, so that there is no relation, for example, between a cat's color and tail length. This is actually only true for genes that are not linked to each other.

Independent assortment occurs during meiosis I in eukaryotic organisms, specifically anaphase I of meiosis, to produce a gamete with a mixture of the organism's maternal and paternal chromosomes. Along with chromosomal crossover, this process aids in increasing genetic diversity by producing novel genetic combinations.

Of the 46 chromosomes in a normal diploid human cell, half are maternally-derived (from the mother's egg) and half are paternally-derived (from the father's sperm). This occurs as sexual reproductioninvolves the fusion of two haploid gametes (the egg and sperm) to produce a new organism having the full complement of chromosomes. During gametogenesis—the production of new gametes by an adult—the normal complement of 46 chromosomes needs to be halved to 23 to ensure that the resulting haploid gamete can join with another gamete to produce a diploid organism. An error in the number of chromosomes, such as those caused by a diploid gamete joining with a haploid gamete, is termed aneuploidy.

In independent assortment the chromosomes that end randomly sorted from all possible combinations of maternal and paternal chromosomes. Because gametes end up with a random mix instead of a pre-defined "set" from either parent, gametes are therefore considered assorted independently. As such, the gamete can end up with any combination of paternal or maternal chromosomes. Any of the possible combinations of gametes formed from maternal and paternal chromosomes will occur with equal frequency. For human gametes, with 23 pairs of chromosomes, the number of possibilities is 2 or 8,388,608 possible combinations. The gametes will normally end up with 23 chromosomes, but the origin of any particular one will be randomly selected from paternal or maternal chromosomes. This contributes to the genetic variability of progeny.

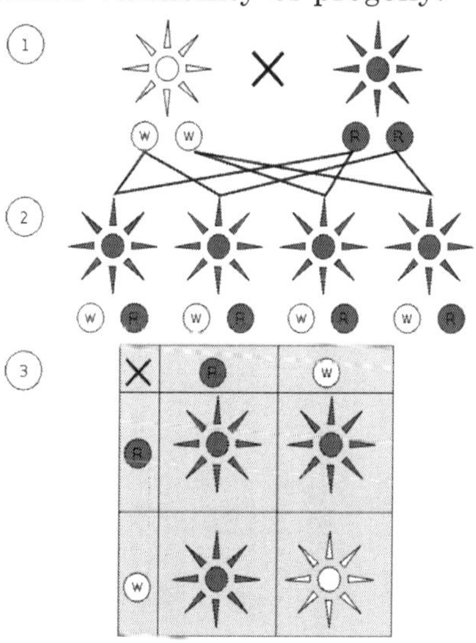

Fig. 5: Dominant and recessive phenotypes.

(1) Parental generation. (2) F_1 generation. (3) F_2 generation. Dominant (red) and recessive (white) phenotype look alike in the F_1 (first) generation and show a 3:1 ratio in the F_2(second) generation

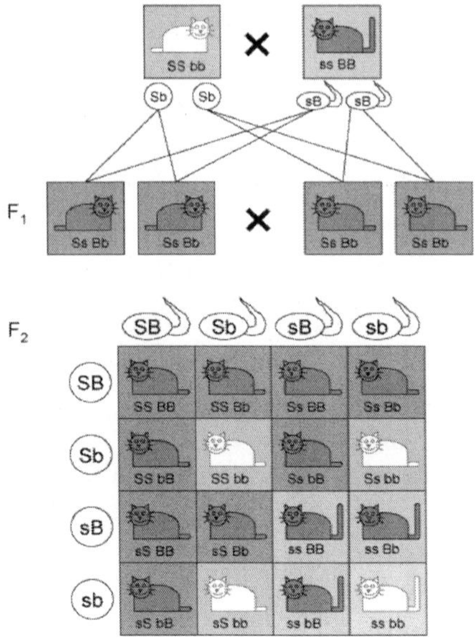

Fig. 6: The phenotypes of two independent traits show a 9:3:3:1 ratio in the F_2 generation. In this example, coat color is indicated by B (brown, dominant) or b(white) while tail length is indicated by S (short, dominant) or s (long). When parents are homozygous for each trait ('SSbb and ssBB), their children in the F_1generation are heterozygous at both loci and only show the dominant phenotypes. If the children mate with each other, in the F_2 generation all combination of coat color and tail length occur: 9 are brown/short (purple boxes), 3 are white/short (pink boxes), 3 are brown/long (blue boxes) and 1 is white/long (green box).

Background

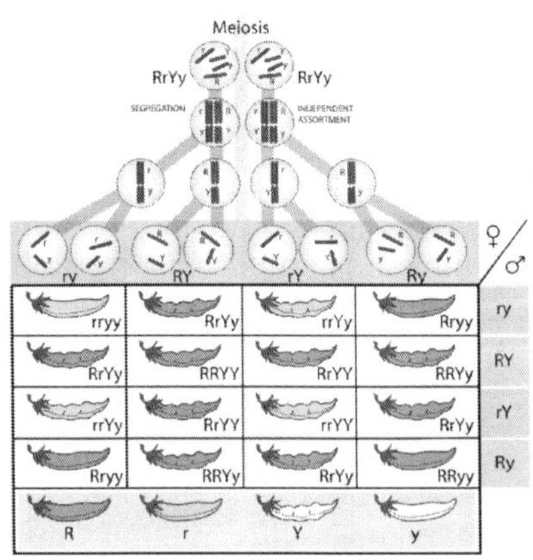

Fig. 7: Table showing how the genes exchange according to segregation or independent assortment during meiosis and how this translates into Mendel's laws

Fig. 8: Mendel's law: Forget-me-not

The reason for these laws is found in the nature of the cell nucleus. It is made up of several chromosomes carrying the genetic traits. In a normal cell, each of these chromosomes has two parts, the chromatids. A reproductive cell, which is created in a process called meiosis, usually contains only one of those chromatids of each chromosome. By merging two of these cells (usually one male and one female), the full set is restored and the genes are mixed. The resulting cell becomes a new embryo.

The fact that this new life has half the genes of each parent (23 from mother, 23 from father for total of 46 in the case of humans) is one reason for the Mendelian laws. The second most important reason is the varying dominance of different genes, causing some traits to appear unevenly instead of averaging out (whereby dominant doesn't mean more likely to reproduce—recessive genes can become the most common, too).

There are several advantages of this method (sexual reproduction) over reproduction without genetic exchange:

1. Instead of nearly identical copies of an organism, a broad range of offspring develops, allowing more different abilities and evolutionary strategies.
2. There are usually some errors in every cell nucleus. Copying the genes usually adds more of them. By distributing them randomly over different chromosomes and mixing the genes, such errors will be distributed unevenly over the different

children. Some of them will therefore have only very few such problems. This helps reduce problems with copying errors somewhat.

3. Genes can spread faster from one part of a population to another. This is for instance useful if there's a temporary isolation of two groups. New genes developing in each of the populations don't get reduced to half when one side replaces the other, they mix and form a population with the advantages of both sides.
4. Sometimes, a mutation can have positive side effects. For example, sickle cell anemia is a mutation that can cause the benefit of malaria resistance. The mechanism behind the Mendelian laws can make it possible for some offspring to carry the advantages without the disadvantages until furthermutations solve the problems.

Mendelian Trait

A Mendelian trait is one that is controlled by a single locus and shows a simple Mendelian inheritance pattern. In such cases, a mutation in a single gene can cause a disease that is inherited according to Mendel's laws. Examples include sickle-cell anemia, Tay-Sachs disease, cystic fibrosis and xeroderma pigmentosa. A disease controlled by a single gene contrasts with a multi-factorial disease, like arthritis, which is affected by several loci (and the environment) as well as those diseases inherited in a non-Mendelian fashion. The Mendelian Inheritance in Man database is a catalog of, among other things, genes in which Mendelian traits cause disease.

Chapter 7

DNA Quantification and Quality Assessment

DNA quantification and quality assessment is done by using UV-VISspectrophotometry.Normally quality check is performed through the A260/A280 ratio that is 1.8 value shows the highest purity, if more than 1.8 shows the presence of RNA contamination and less than that shows the protein contamination (Sambrook, *et. al,* 1989).

AGE is conducted of isolated DNA samples for visualization of bands. For perfect extraction, there will be clear single bands otherwise smear like image will be seen. That is in case of degraded DNA. For PCR based methods the DNA samples are diluted for PCR requirements. The optimum concentration is usually 40- 60ng/μL.

Polymerase Chain Reaction (PCR)

The DNA amplification by thermal cycling called Polymerase Chain Reaction is in vitro method that can be used to amplify a specific DNA segment from small amounts of DNA template or duplex into millions of copies. It is invented by Kary Mullis *et.al,* (1985).

Heatdenaturation

This temperature denatures the double stranded DNA into two individual stands. Denaturation temperature is 95^0 C for 30 seconds or is 97^0 C for 15 seconds, however higher temperature may be appropriate, especially for G+C rich nucleotides.

Annealing

During this time one primer binds with the 5 prime end of one DNA strand and the other primer binds with 3 primo end of its complementary strand .Annealing is hybridization of primers to single stranded DNA and the length of time required for primer annealing depends on the basic composition, length and concentration of primers.

Primer Extension

This temperature is varies for Taq DNA polymerase which adds complementary nucleotides one by one to the 3' OH group of the primer. Estimates for the rate of nucleotide incorporation at 72^0 C vary from 35-100 nucleotides per second depending upon the buffer, pH salt concentration and nature of DNA template.

Number of PCR Cycles

The optimum number of cycles depends mainly upon the starting concentration of target DNA, when other parameters are optimized. A common mistake is to execute too many cycles. The increase in the number of cycles will increase the amount and complexity of non-specific background products.

Restriction Fragment Length Polymorphisms (RFLPs)

RFLP definition: The variation(s) in the length of DNA fragments produced by a specific restriction endonuclease from genomic DNAs of two or more individuals of a species (Kahl, 2001).

Restriction fragment length polymorphisms (RFLPs) was the first to develop (Soller and Beckmann, 1983) and are still widely used marker system especially in practical plant breeding. It was first developed in the 1980s for use in human genetic applications and was later applied to plants Most of the studies in systematic and evolution (Whitkus *et al.*, 1994) and molecular maps of different crop species were initially made through the use of RFLP markers. Restriction fragment length polymorphism (RFLP) technology. RFLPs are relatively small in size and are co-dominant in nature. Restriction fragments of different lengths are thus generated. All RFLP markers are analysed using a common technique. However, the analysis requires a relatively complex technique that is time consuming and expensive.

The hybridization results can be visualized by autoradiography (if the probes are radioactively labelled), or using chemiluminesence (if non-radioactive, enzyme-linked methods are used for probe labelling and detection). Any of the visualization techniques will give the same results. The visualization techniques used will depend on the laboratory conditions. A comprehensive study on these maps is documented by Phillips and Vasil (1994). Thus, RFLP markers have the potential to address with power, precision and accuracy, all issues of plant breeding in which markers can be applied. In addition to that, voluminous literature is available on utilization of RFLPs in other aspects especially in DNA fingerprinting (Tanksley, 1983), phylogenetic analysis (Zheng *et al.*, 1994), detection of genetic variability in rice, genetic dissection of maturity (Phillips *et al.*, 1992), identification of chromosomal regions responsible for regeneration of RFLPs, comparative mapping of chromosomes in wheat, rye and barley, analysis of wide compatibility in rice varieties (Zheng *et al.*, 1994), QTL mapping, gene tagging, and identification of ecotypes (Pupilli *et al.*, 2000).

To discover RFLPs, restriction enzymes (RE) are used to cut DNA at specific 4-6 bp recognition sites. Sample DNA is digested with one or more RE's and resulting fragments are separated according to molecular size using gel electrophoresis. After that Southern Blotting is performed where fragments are immobilized on a nylon membrane, followed by hybridization to a labelled probe (specific chemoluminescence DNA probe) that identifies the locus under autoradiograph. Fragment sizes of digested DNA will differ depending on the presence or absence of the proper recognition sequence for the enzyme.

Differences in fragment length result from base substitutions, additions, deletions or sequence rearrangements within RE recognition sequences. Although two individuals of the same species have almost identical genomes, they will always differ at a few nucleotides. Some of these differences will produce new restriction sites (or remove them), and the banding pattern seen on a genomic Southern will thus be affected. For any given probe (or gene), it is often possible to test different restriction enzymes until one which gives a pattern difference between two individuals is found, that is, a RFLP. The less related the individuals, the more divergent their DNA sequences are and the more likely you are to find a RFLP.

However, the low level of polymorphism present in RFLPs restricted the use of this technique. In addition, RFLP requires large quantities of relatively pure DNA, and the frequent use of radioisotopes in detection step makes it technically demanding, laborious and costly to characterize large number of samples.

Utilization of RFLP markers requires a lot of skill in using short-lived radioactive material: the availability of which often becomes a limiting factor especially in the developing countries. Another limiting factor could be the large sample size to be analysed like that of F segregating population. In this case, utilization of RFLP markers may not be feasible, as it requires extraction of DNA in ultra pure form, restriction digestion, and lengthy and expensive process of isotopic labeling. Such difficulties have opened up a debate on the feasibility of routine application of RFLPs in large scale crop improvement programme.

Another difficulty frequently encountered by RFLP users is the detection of low (1-3 loci) ornon-existent polymorphism in closely related species/varieties (Waugh and Powell, 1992). It was essentially this difficulty that led to the search for a new marker system.

Randomly Amplified Polymorphic DNAs (RAPDs)

Randomly-amplified polymorphic DNA markers (RAPD). In 1991 Welsh and McClelland developed a new PCR-based genetic assay namely randomly amplified polymorphic DNA (RAPD). This procedure detects nucleotide sequence polymorphisms in DNA by using a single primer of arbitrary nucleotide sequence.

In this reaction, a single species of primer anneals to the genomic DNA at two different sites on complementary strands of DNA template. If these priming sites are within an amplifiable range of each other, a discrete DNA product is formed through thermocyclic amplification. On an average, each primer directs amplification of several discrete loci in the genome, making the assay useful for efficient screening of nucleotide sequence polymorphism between individuals.

However, due to the stoichastic nature of DNA amplification with random sequence primers, it is important to optimize and maintain consistent reaction conditions for reproducible DNA amplification. RAPDs are DNA fragments amplified by the Polymerase Chain Reaction (PCR) using short (generally 10 bp) synthetic primers of random sequence.

Because of the short primers used, the reannealing temperature in the PCR must be low (35-40°C) for the primer to bind. However, due to low temperature, the binding is not very specific, which means that primers will bind also to sequences which are not completely complementary. These oligonucleotides serve as both forward and reverse primer and usually are able to amplify fragments from 3-10 genomic sites simultaneously. Amplified fragments (within the 0.5-5 kb range) are separated by gel-electrophoresis and polymorphisms are detected as the presence or absence of bands of particular size. These polymorphisms are considered to be primarily due to variation in the primer annealing sites.

They are dominant markers and hence have limitations in their use as markers for mapping, which can be overcome to some extent by selecting those markers that are linked in coupling.

RAPD assay has been used by several groups as efficient tools for identification of markers linked to agronomically important traits, which are introgressed during the development of near isogenic lines. The application of RAPDs and their related modified markers in variability analysis and individual-specific genotyping has largely been carried out, but is less popular due to problems such as poor reproducibility faint or fuzzy products, and difficulty in scoring bands, which lead to inappropriate inferences.

In RAPD analysis the total DNA from an organism is mixed with (generally) 10 bp single stranded DNA (primer) together with the four different deoxynucleotides and a heat stable DNA polymerase enzyme. The reaction mixture is placed on a thermocycler (PCR-machine) which can change the temperature of the reaction rapidly according to a predefined PCR programme.

Reannealing at low temperature makes the primer molecules attach to complementary site on the DNA. Elongation is the period when the polymerase enzyme elongates the attached primers from their 3' end. Denaturation separates the new formed strands.

After a few PCR cycles DNA pieces with well defined length similar to the one between the two original primer sites dominate the mixture. This is because they amplify exponentially. In most RAPD reactions the result will be amplification of 5 to 10 different DNA segments whose lengths depend on the primer recognition sites available in the genome. The DNA after amplification is visualized for bands using Agarose Gel Electrophoresis (AGE). The band scoring is done by comparing with the marker bands.

Inter Simple Sequence Repeats (ISSR)

ISSR amplification definition: A variant of the polymerase chain reaction that uses simple sequence repeat primers (e.g. [AC] n) to amplify regions between their target sequences (Kahl, 2001).

Inter simple sequence repeat markers (ISSR). In this technique, reported by Zietkiewicz et al. primers based on microsatellites are utilized to amplify inter-SSR DNA sequences. Here, a single primer or combination of primers composed of a microsatellites sequence anchored with a single base at 3' region so that it can amplify regions between two microsatellites loci with oligonucleotides annealing to repeat sequence. More complex banding patterns can be achieved using 5'-anchored primers that incorporate the SSR regions in their amplification products, and by combining 3'- and 5'- primers.ISSR circumvents the requirement for flanking sequence information, and thus has found wide applicability in a variety of plants.(Wolfe et al.1998} These are mostly dominant markers, though occasionally a few of them exhibit codominance. An unlimited number of primers can be synthesized for various combinations of di-, tri-, tetra- and pentanucleotides [(4)3 = 64, (4)4 = 256] etc. with an anchor made up of a few bases and can be exploited for a broad range of applications in plant species. ISSR analysis is technically simpler than any other marker systems. The method provides highly reproducible results and generates higher levels of polymorphism compared to other markers in many systems. The advantage lies in long primer length and stringent annealing temperature.

Primers used (from Yang *et al.,* 1996):

For separation and visualization of bands Agarose gel electrophoresis is conducted.

Simple Sequence Repeats (SSRs)

SSR (Microsatellites) definition: Any one of a series of very short (2-10 bp), middle repetitive, tandemly arranged, highly variable (hyper variable) DNA sequences dispersed throughout fungal, plant, animal and human genomes (Kahl, 2001).

Simple sequence repeats (SSR) or microsatellites are a class of repetitive DNA elements. The di-, tri- or tetra-nucleotide repeats are arranged in tandem arrays consisting of 5 – 50 copies, such as (AT) 29, (CAC) 16 or (GACA) 32. SSRs are abundant in plants, occurring on average every 6-7 kb (Cardle *et al.*, 2000). These repeat motifs are flanked by conserved nucleotide sequences from which forward and reverse primers can be designed to PCR-amplify the DNA section containing the SSR.SSR alleles, amplified products of variable length, can be separated by gel electrophoresis and visualized by silver-staining, autoradiography (if primers are radioactively labelled) or via automation (if primers are fluorescently labelled). SSR analysis is amenable to automation and multiplexing, and allows genotyping to be performed on large numbers of lines, and multiple loci to be analysed simultaneously. SSRs can be identified by searching among DNA databases (e.g. EMBL and Gene bank), or alternatively small insert (200-600bp) genomic DNA libraries can be produced and enriched for particular repeats (Powell *et al.*, 1996). From the sequence data, primer pairs (of about 20 bp each) can be designed (software programmes are available for this).

For separation and visualization of bands Polyacrylamide Gel Electrophoresis (PAGE) is conducted (6% acrylamide solution), stained and the observed the bands.

Amplified Length Polymorphisms (AFLPs)

AFLP definition: Any difference between corresponding DNA fragments from two organisms A and B that is detected by the amplified restriction length polymorphism technique (Kahl, 2001).

Table 1: DNA Fingerprinting Techniques: A Comparative Analysis

Marker	*Advantages*	*Disadvantages*
RFLP	Unlimited number of loci	Labour intensive
	Co-dominant	Fairly expensive
	Many detection systems	Large quantity of DNA needed
	Can be converted to SCARs	Often very low levels of polymorphism
	Robust in usage	Can be slow (often longexposure times)
	Good use of probes fromother species	Needs considerable degree ofskill
	Detects in related genomes	
	No sequence informationrequired	
RAPD	Results obtained quickly	Highly sensitive to laboratorychanges
	Fairly cheap	Low reproducibility within andbetween laboratories
	No sequence informationrequired	Cannot be used acrosspopulations nor across species
	Relatively small DNAquantities required	Often see multiple loci
	High genomic abundance	Dominant
	Good polymorphism	
	Can be automated	

ISSR	Highly polymorphic	Usually dominant
	Robust in usage	Species-specific
	Can be automated	
SSR	Fast	High developmental andstartup costs
	Highly polymorphic	Species-specific
	Robust	Sometimes difficult interpretation because of stuttering
	Can be automated	Usually single loci even in polyploids
	Only very small DNA	<!—[if !support Misaligned Columns]—>
		<!—[endif]—>
AFLP	Small DNA quantitiesrequired	Evaluation of up to 100 loci
	No sequence informationrequired	Marker clustering
	Can be automated	Dominant
	Can be adapted for differentuses, e.g.	Technique is patented
cDNA-AFLP		Can be technically challenging

This genetic marker system basically uses restriction fragment analysis, but a PCR approach is used to visualize fragments instead of hybridization with a probe. Because of the PCR step, this type of markers has characteristics quite different from RFLP markers. Initially, DNA is extracted as for RFLP and then cut with two different restriction enzymes to produce well defined restriction fragments with sticky ends. Synthetic double stranded linkers of approximately 18 to 20 bp with matching sticky ends are ligated on all the restriction fragments. Ligated fragments are subsequently amplified in PCR with 18 to 20 nucleotide length primers recognizing linkers in each end of the fragments. A primer used for amplification in AFLP is normally labelled with P^{33} or some nonradioactive labeling system. The mixture of amplified fragments are separated according to size in a Polyacrylamide gel and visualized by means of autoradiography or other procedures for development of non radioactive systems. Alternatively, automated sequencing machines can be used to read the sequence of amplification products directly from the Polyacrylamide gel if primers have been labelled with special fluorescent dyes. Also, silver staining of the Polyacrylamide gel after electrophoresis can be used to stain all amplified DNA fragments in the gel. To avoid that all restriction fragments from the genome amplify in the same PCR, which will produce a smear because of too many DNA fragments, a two step amplification procedure is used. Firstly, part of the total number of restriction fragments are amplified in the "preamplification" with primers containing one extra "selective" nucleotide on their 3' end.

This selective nucleotide will allow amplification only of restriction fragments with a matching nucleotide next to the linker. With one selective nucleotide on both primers only 1/16th of all restriction fragments in the mixture will amplify during the preamplification. During the subsequent "selective" amplification, additionally, one or two selective nucleotides on each primer will further reduce the number of fragments amplified. A good AFLP amplification will show 90 to 100 different fragments in one analysis.

Applications

There are a variety of applications in DNA fingerprinting, such as to identify genetic diversity within breeding populations in plants and animals, to differentiate between plant species cultivars, as well as to identify plants containing a gene of interest. , it is also used to detect genetically modified organisms in agriculture in the United States.

RFLP is most suited to studies at the intraspecific level or among closely related taxa. Presence and absence of fragments resulting from changes in recognition sites are used for identifying species or populations. . In plant genetic analysis, RFLP markers were initially used for estimating genetic distance and fingerprinting in wheat. In plant and animal breeding, they can reveal which backcross progeny most resemble the desired parent. RFLPs are used to settle cases where biological parentage is at issue.

Many plant species are being analysed using RAPD method. RAPD markers were also used for characterisation, estimation of genetic relatedness and determination of genetic diversity of tea germplasm. These markers are able to distinguish between genotypes but limited to comparisons of populations from a few sources. Lastly RAPD analysis was used to assess genetic differences and relatedness among fig, *Ficus carica* L. clones.

For AFLP analysis, it is an efficient technology for assessing genetic diversity among cultivars such as wheat, *Triticum aestivum* cultivars. . Cultivars were improved through crossing with elite breed however, crossing within such germplasm may lead to a reduction in the level of genetic variation among parental lines due to intense selection pressure. Thus DNA marker systems are used to reflect actual DNA differences among lines since selection pressure and genetic drift are accounted for. It is an efficient genetic marker system as the technique gives a much higher degree of polymorphism than RAPD or RFLP. In addition, its prior sequence knowledge is not required as the fragments detected are selected by specific primer sets in the PCR reaction. It is reliable as stringent reaction conditions are applied for primer annealing.

Inter SSR (ISSR) fingerprinting was developed such that no sequence knowledge was required. Primers based on a repeat sequence, such as (CA) n, can be made with a degenerate 3'-anchor, such as (CA) 8RG or (AGC) 6TY. The resultant PCR reaction amplifies the sequence between two SSRs, yielding a multilocus marker system useful for fingerprinting, diversity analysis and genome mapping. Previous investigators have demonstrated that ISSR analysis usually detects a higher level of polymorphism than that detected with restriction fragment length polymorphism (RFLP) or random amplified polymorphic DNA (RAPD) analyses.

As with AFLPs, the great advantage of SSRs analysis is the large number of polymorphisms that the method reveals, with one locus in soybean (Glycine max) reported to have 26 alleles. Unlike AFLPs, microsatellites are co-dominant markers, thus heterozygotes can be readily identified. Microsatellite co-dominance will increase the efficiency and accuracy of population genetic measures based on these markers compared with other markers, such as AFLPs and RAPDs.

With a beginning as a mere forensic tool, the world of DNA fingerprinting has a gone a long way in revealing the genetic identity of living beings. With massive evolution in concepts and techniques it has given a lot to various fields cutting across the spectrum. In plants it has not only helped in identifying species but also in defining a new realm in plant genomics, plant breeding and in conserving the biodiversity. With world paving way for developments in biotechnology, DNA fingerprinting promises a very powerful tool in our future endeavors.

Intellectual Property Rights in Plant Varieties

Legal and Policy Concepts Relating to the Protection of Plant Varieties and Plant Breeders' Rights

Intellectual property rights (IPRs) are legal rights granted by governmental authorities to control certain products of human intellectual effort and ingenuity. (OECD 1996). An in-depth discussion of the philosophical and policy goals served by granting legal protection to these products is beyond the scope of this study. However, a basic familiarity with these goals is necessary to grasp how national and international intellectual property systems and institutions have evolved into their present forms and to understand the constraints that those systems and institutions place on governments seeking to implement competing policy objectives in tension with IPRs.

Two broad philosophical approaches underlie the decision to grant IPRs in the products of human intellectual effort and ingenuity. Elements of these two approaches can be found to different degrees in all national laws and international agreements relating to IPRs.

The first approach to IPR protection predominates in many civil law legal systems, including continental Europe. It takes the position that the products of the human mind are stamped with the personality of their creator, inventor or author, thus endowing him or her with a moral as well as an economic claim to exploit those products to the exclusion of third parties. Under this view, legal protection flows from a state's commitment to protect human rights, a fact reflected in the wording of article 27 of the Universal Declaration of Human Rights, which guarantees to everyone "the right to the protection of the moral and material interests resulting from any scientific, literary or artistic production of which he is the author." Legal protection also rewards creators, inventors and authors for their intellectual efforts as well as their expenditures of time and money. From these underlying premises flows a desire to provide robust intellectual property protection that includes, for example, a broad and expansive range of exclusive rights, long terms of protection, limits on mandatory licensing and narrow exceptions and limitations to exclusive rights.

The second approach to IPR protection takes as its starting premise an instrumental view of intellectual property. Legal protection for the products of human intellectual effort and ingenuity is granted not because of a moral commitment to compensating creators or innovators, but rather because the products they create enrich a society's culture and knowledge and thus increase its welfare. Perhaps the most well-known manifestation of this approach is found in the Intellectual Property Clause of the United States Constitution, which authorizes the United States Congress "[t]o promote the progress of science and useful arts, by securing for limited times to authors and inventors the exclusive right to their respective writings and discoveries."

This instrumentalist philosophy shapes the structure of many national intellectual property systems. The grant of IPRs in nations following an instrumentalist approach is intended to provide adequate incentives for creators, inventors and authors to invest the time, resources and intellectual capital needed to create intellectual property products.

In the absence of a grant of exclusive rights over those products, so-called "free riders" who have not made such investments could exploit inexpensive distribution and reproduction technologies and sell others' intellectual property products at a much lower cost. However, the ultimate goal of legal protection is not remunerative reward for creators but the enhancement of social welfare through access to the ideas and information contained in their products. Instrumentalist intellectual property systems often tailor the scope of legal protection to achieve this goal, for example by placing certain limits on the scope of protection or recognizing situations in which consumers or second-generation creators may access and use intellectual property products for socially valuable purposes.

Policy Objectives Favouring IPRs in New Plant Varieties

The policy goals of granting IPRs to plant varieties are grounded principally on an instrumentalist approach to IPRs. This is true both for patents and plant breeders' rights.

Under this instrumentalist approach, new plant varieties are afforded legal protection to encourage commercial plant breeders to invest the resources, labour and time needed to improve existing plant varieties by ensuring that breeders receive adequate remuneration when they market the propagating material of those improved varieties. In the absence of a grant of exclusive rights to breeders, the dangers of free riding by third parties would be considerable. This is because the genetic material within plants that specifies their distinctive and commercially valuable features is naturally self-replicating, for example by reproduction of seeds or other propagating material. Self-replication makes innovations incorporating biological material particularly susceptible to exploitation by parties other than the innovator.

IPRs in plant varieties thus provide some assurance to breeders that they will be able to recoup the risks and costs of a value-added innovation that is based upon an underlying biological resource.

Ultimately, however, the grant of exclusive rights to plant breeders is designed to benefit the society granting the rights. It provides an incentive for private research and development into new breeding techniques, thereby reducing the need for government funding to subsidize these activities. It encourages the development of new and beneficial plant varieties for use by farmers and consumers. And it furthers the society's development of agriculture, horticulture and forestry.

An international system of IPR protection for plant varieties expands these benefits by facilitating access to new varieties created in other states. Once breeders are assured that their rights will be protected in other states, breeders will be more willing to make their new varieties available in those states (assuming they have access to a distribution and marketing infrastructure). This benefits farmers, consumers and researchers in many more jurisdictions.

The Evolution and Structure of the International Intellectual Property System

The different policy objectives underlying the protection of IPRs have shaped the structure and evolution of the international intellectual property system. Most early domestic intellectual property laws provided no legal protection to intellectual property products created in other nations, thereby permitting those products to be exploited by free riders operating outside the state in which the products were created. The unfairness of this result prompted governments in the late nineteenth century to consider an international approach to protect IPRs.

Limited treaty obligations: The drafters of the first multilateral intellectual property treaties quickly realized, however, that there was insufficient political support for reconciling many of the differences that existed among national IPR laws. For this reason, the drafters abandoned the idea of harmonizing diverse national laws to create a single, international IPR applicable in all signatory states. They fashioned instead a system that creates a limited set of treaty-based obligations that each member state of that system is then required to implement in its national IPR laws.

Implementation of treaty-based obligations in national IPR laws can occur in one of two ways. In some nations (often referred to as "automatic incorporation" states), treaties become binding as a part of domestic law as soon as formal ratification procedures have been adopted. In these nations, treaties are considered to be "self-executing" or capable of being given "direct effect" in domestic law such that courts and administrative agencies can construe the treaty directly and enforce the rights it grants to the owners of intellectual property products.

In other nations, however (often referred to as "legislative incorporation" states), treaties are considered to be "non-self-executing" and can only become binding in domestic law once the parliament or legislature has adopted legislation to implement the treaty. In these nations, owners of intellectual property products rely on this domestic legislation rather than on the treaties themselves when they seek to enforce rights granted to them under the treaties.

The territoriality of IPRs : Because of the limited scope of international IPR agreements, there are to this day (with the limited exception of the European Union) no *international* IPRs available to inventors and creators who seek to market their products across borders; rather, IPRs are territorial in nature and are acquired and enforced on a country-by-country basis under territorially-circumscribed *national* IPR laws.

Thus, for example, the inventor of a genetically enhanced variety of corn who seeks patent protection for that variety must apply for protection in each country in which he or she hopes to sell the corn. The inventor must comply with all of the requirements that each country imposes for granting patent rights to the new variety. Similarly, once protection is granted, issues such as the scope of the exclusive rights the inventor enjoys in the patented plant variety, the term of patent protection and the limitations imposed on the inventor's rights are all determined by the different national laws.

Recent international agreements have achieved some modest forms of procedural harmonization, but they have not altered the fundamental premise that national laws rather than international treaties are the immediate source of nearly all private rights in intellectual property products.

Two basic principles flow from this territorial approach to protecting IPRs. First, where national laws differ as to the scope or content of the protection they provide to intellectual property products, the rights enjoyed by the owners of those products will vary in different national jurisdictions. Second, territoriality implies that each nation has the right to decide on the form of IPR protection to be granted within its own borders, provided that it complies with the obligations contained in international IPR agreements to which it is a party.

Although territoriality thus gives governments some autonomy to set national IPR policies within their own borders, states often view the policies other governments choose as a subject of concern. Indeed, the global reach of markets for intellectual property products makes this concern a necessity.

To take just one example, most patent laws grant inventors owning patents within a state the right to prevent the importation into that state of products created in other nations that contain the patented invention. Thus, where distribution markets transcend national borders, an industry may find itself precluded from distributing products in other jurisdictions as a result of patent rights, as occurred when Indian cotton producers were precluded from importing certain forms of transgenic cotton into the United States.

Core obligations imposed by international intellectual property agreements : The territorial approach to IPR protection appears at first to present myriad difficulties for creators and owners of intellectual property products. In fact, however, the content of each nation's IPR laws are often quite similar since they have been shaped by international IPR agreements ratified by many states. In addition, the obligations these agreements impose have expanded over time, thus narrowing (although by no means eliminating) the differences among national intellectual property systems. The following sections briefly explain the core obligations contained in most international IPR agreements. More detailed information concerning the rights and obligations contained in specific IPR agreements relating to plant varieties is provided in Part II below.

National treatment : One of the cornerstones of international IPR agreements is the national treatment principle. National treatment bars discrimination against foreign IPR owners by requiring that each state provide the same IPRs to private parties from other member states as are provided to the state's own nationals. National treatment levels the playing field among treaty parties and prevents a state from giving its own creators and inventors unfair advantages over foreign creators and inventors. In the absence of national treatment, for example, domestic firms could freely exploit intellectual property products created in other member states while simultaneously enjoying legal protection within their own domestic markets.

Reciprocity : The provisions of several intellectual property treaties contain a limited exception to national treatment known as reciprocity. Where a treaty permits reciprocity, State A may condition the grant of legal protection to intellectual property products from State B upon State B's granting of legal protection to intellectual property products from State A. Reciprocity is often applied to new IPRs as means of encouraging other nations to recognize the new rights and extend their protection to foreign nationals. Once a large number of states have recognized the new IPR, they may revise the treaty to eliminate the reciprocity option and impose a national treatment obligation.

Most favoured nation treatment : The most favoured nation ("MFN") principle is a common feature of international trade agreements but has only recently been applied to IPRs. The principle extends the national treatment rule by compelling a government that provides a privilege or benefit to *one* state within a treaty system automatically to grant that same privilege or benefit to *all* states within the same system. The MFN principle thus prevents a subset of states within a larger treaty system from entering into bilateral or other special agreements among themselves, unless they grant the rights contained in those agreements to all other parties within the larger treaty system.

Subject matter and eligibility requirements : Intellectual property agreements specify the subject matter characteristics of the products that are eligible for legal protection. In the context of patents, for example, a treaty may specify the types of inventions (such as products and processes) to which states parties must grant legal rights. These subject matter requirements are generally drafted using language that instructs member states concerning the basic characteristics that a product must possess for it to merit protection under domestic IPR laws, while preserving sufficient flexibility for states to tailor the details of protection to the particularities of their national legal systems.

Exclusive rights : Where an intellectual property product satisfies a treaty's subject matter eligibility requirements, states parties are required to grant an enumerated set of exclusive rights with respect to that product. These exclusive rights grant to the product's owner the power to exclude all third parties from engaging in the activity that the right covers (such as reproducing or modifying the product or distributing it to others). It is the exclusivity of the rights granted that allows IPR owners to recoup the investment of time, money and resources required to create intellectual property products. The particular exclusive rights mandated by IPR agreements differ depending on the specific type of IPR product at issue. They are drafted with greater or lesser degrees of specificity depending on the degree of consensus among member states' national IPR laws, thus affording varying levels of discretion to governments to implement the rights in their domestic legal systems.

Terms of protection and the public domain : Intellectual property agreements also specify for their states parties the minimum term of years during which intellectual property products must receive legal protection. Once that term has expired, the treaties do not require states to grant legal protection to the products. Thus, unless the state adopts a longer term of protection, after the expiration of the initial term of protection the product may be freely used by anyone for any purpose, including as a source for creating new products or simply for consumption. A corollary of this rule is that national IPR laws do not permit putative inventors and creators to claim IPRs in materials as they are found in nature or where they are already part of the public domain.

Exceptions and limitations to exclusive rights : International IPR agreements constrain the ability of national governments seeking to restrict the exercise of IPRs to achieve competing social or policy objectives, such as access to information, research, education and cultural development. Restrictions designed to achieve these objectives are generally known as "exceptions and limitations" to exclusive IPRs. These exceptions and limitations generally appear in two forms. The first form permits third parties to engage in specified uses of intellectual property products without the permission of the rights holder and without the payment of remuneration. The second form is known as a "compulsory licence." Compulsory licences allow third parties to use intellectual property products without the owner's permission, but only upon the payment of adequate compensation. To prevent both forms of exceptions and limitations from eviscerating IPRs altogether, intellectual property agreements impose specific constraints the ability of member states to adopt them.

Enforcement provisions : The grant of IPRs in national laws would be meaningless without adequate and effective mechanisms to enforce those rights. For this reason, recent intellectual property agreements specify the types of enforcement provisions that member states must adopt in their national laws. These provisions include the imposition of civil and criminal penalties against any person who engages in acts of exploitation reserved to the owner of an intellectual property product without the owner's authorization. The penalties include civil judicial proceedings for monetary damages or an injunction to prevent the continued unauthorized use of the product and criminal proceedings commenced by the government itself.

The "Minimum Standards" Framework of Multilateral IPR Agreements

Taken together, these core provisions of international IPR agreements impose significant legal obligations on member states. The agreements do not, however, purport to definitively address all of the issues raised by the grant of legal protection to intellectual property products. For this reason, the treaties are often referred to as "minimum standards" agreements in that they create only a basic floor of legal protection to which all member states must adhere.

There are three important consequences of this minimum standards framework. First, it allows member states the discretion to interpret and apply those provisions of the treaties that are ambiguous or that reasonably permit more than one construction. Second, a minimum standards approach permits, but does not require, states to grant additional IPR protections within their national laws. And third and perhaps most importantly, the framework leaves member states free to enact laws that serve other political, economic or social objectives, even where those objectives are in tension with IPRs, provided that those laws are not inconsistent with the terms of IPR agreements.

In the last five years, however, an increasing number of developing countries have entered into bilateral or regional trade and investment treaties with the United States or with the European Community. These treaties often contain higher standards of intellectual property protection than those specified in multilateral intellectual property agreements. (Drahos, 2001; GRAIN, 2001) They thus restrict the minimum standards approach contained in the multilateral agreements, for example, by requiring states to enact specific intellectual property protection rules in their national laws or by requiring developing countries to follow the standards contained in multilateral agreements to which they are not parties.

The minimum standards framework (as limited by bilateral or regional treaties requiring higher standards of protection) provides a methodology for analysis but not a solution for every potential clash between IPRs and other governmental policies. This is particularly true for plant varieties and plant breeders' rights, an area of intellectual property protection regulated by several international IPR agreements and subject to diverse standards of legal protection under different domestic laws. To understand how national governments can reconcile competing and sometimes conflicting domestic laws and objectives consistently with the obligations imposed by different international IPR agreements, it is first necessary to examine other international agreements and international institutions relating to plant genetic resources that have promulgated policies in tension with IPRs.

Identifying the relevant international agreements and institutions relating to intellectual property rights in plant varieties.

Before turning to a more detailed discussion of policy objectives, it is important first to identify the principal international institutions and international agreements that generate legal rules and standards relating to IPRs in plant varieties and plant genetic resources, as well as the institutions and agreements that generate rules and standards in tension with IPRs. Although a comprehensive discussion of these institutions and agreements is beyond the scope of this study, a basic familiarity with their most important elements is necessary to understand the legal regime of intellectual property rights in plant genetic resources for food and agriculture.

World Intellectual Property Organization ("WIPO")

The World Intellectual Property Organization ("WIPO") is a specialized agency of the United Nations charged with "promot[ing] the protection of intellectual property throughout the world." (WIPO Convention, art. 3(i)). The WIPO Secretariat undertakes wide variety of activities relating to IPRs, including hosting diplomatic conferences of government representatives seeking to negotiate new international treaties. WIPO's staff also provide technical assistance and training to member states and their national intellectual property offices, especially in developing countries. More recently, WIPO has created standing, expert and intergovernmental committees that conduct studies on particular intellectual property topics and generate nonbinding guidelines and recommendations for consideration by WIPO members.

WIPO's recent activities in the area of plant genetic resources have been considerable. Issues relating to the intersection of IPRs and PGRs have been raised during the negotiation of two multilateral patent agreements (the Patent Law Treaty and the Substantive Patent Law Treaty) and have been a principal subject of discussion and study in the recently created Intergovernmental Committee on Intellectual Property and Genetic Resources, Traditional Knowledge and Folklore ("IGC"). Developing states first sought to raise issues relating to the intersection of IPRs and PGRs during the WIPO-sponsored negotiation of the Patent Law Treaty in 1999.

They proposed the addition of an article in the treaty requiring applicants for inventions derived from genetic resources to demonstrate that they had received permission to access those resources from the country of origin. Industrialized countries opposed the proposal, arguing that it addressed substantive law issues that were inappropriate for inclusion in a treaty largely devoted to procedural issues. As a compromise, the WIPO Secretariat proposed the creation of a new intergovernmental committee (the IGC) to study the intellectual property aspects of genetic resources and traditional knowledge.

During the IGC's first five sessions between April 2001 and July 2003, WIPO members have supported a wide-ranging work programme for the Committee, including: (1) creating a searchable database of clauses in contracts that regulate access to genetic resources or require benefit sharing; (2) studying technical issues raised by the disclosure of biodiversity-related information in patent applications; (3) creating databases of traditional knowledge; (4) identifying ways to document genetic resources and traditional knowledge in the public domain; and (5) studying the appropriate legal rules to protect traditional knowledge.

Most recently, the WIPO General Assembly extended the IGC's mandate and authorized it to accelerate its work, which may include the development of new international agreements.

World Trade Organization ("WTO") : The WTO is a global intergovernmental organization dealing with the rules of trade between nations. It was established in 1994 at the conclusion of the Uruguay Round of trade negotiations held under the auspices of the General Agreement on Tariffs and Trade ("GATT"). As of July 2004, 147 states and customs territories ("WTO Members" or "Members") had joined the organization.

The WTO's substantive obligations are contained in a series of treaties relating to international trade, including agreements on trade in goods, services, agriculture, sanitary and phytosanitary measures, trade-related investment measures, technical barriers to trade, textiles and clothing and intellectual property rights. These treaties are linked together as annexes to the Agreement Establishing the WTO. Disputes between WTO Members relating to any of these treaties are to be resolved by the WTO Dispute Settlement Body, which is composed of *ad hoc* dispute settlement panels and a standing Appellate Body of trade experts. A WTO Member whose national laws or practices are challenged by another WTO Member and found to be incompatible with its WTO treaty obligations by the Dispute Settlement Body must modify those laws or practices or face the prospect of trade sanctions.

A more detailed discussion of the dispute settlement process in the context of the Agreement on Trade Related Aspects of Intellectual Property Rights ("TRIPs" or "TRIPs Agreement"), is provided in paragraph 2.3.1.3.3 below. TRIPs is the WTO agreement that addresses IPRs. It is discussed in detail in section 2.3 below.

Convention on Biological Diversity ("CBD") : The CBD was opened for signature in 1992 and entered into force in 1993. As of July 2004, 188 states had ratified this agreement.

The CBD's main objectives are the conservation of biological diversity, the sustainable use of its components, the fair and equitable sharing of benefits arising out of the utilization of genetic resources and the preservation of indigenous knowledge. (art. 1) The CBD also recognizes that nation states have the sovereign right to exploit their own resources and the authority to determine the conditions of access to them. (arts. 3 and 15)

One of the mechanisms by which the CBD achieves its objectives is *in situ* conversation of plant genetic resources. Conservation *in situ* involves the preservation of ecosystems and natural habitats and the maintenance of viable populations of species in those settings. Such conservation occurs, for example, where farmers and indigenous communities safeguard traditional plant varieties in the locations where they grow naturally or are cultivated.

Although the CBD does not expressly refer to any international IPR agreements, it contains numerous provisions relating to IPRs, principally in article 16. In particular, article 16(5) recognizes that IPRs "may have an influence on the implementation" of the CBD. The article obliges member states to cooperate in order to ensure that IPRs are "supportive of and do not run counter to" the treaty's objectives. Other provisions make clear that the CBD is to be interpreted so as to preserve the rights of IPR owners recognized in international law. For example, articles 16(2)-(4) state that the transfer of technology and measures taken to gain access to such technology shall be consistent with the adequate and effective protection of IPRs recognized in international law. Thus, for example, where a government encourages foreign direct investment in industrial technologies (such as a biotechnological process used to insert new genetic sequences into existing plant varieties), it must respect any patent rights that the owner of that technology has acquired to protect it.

Over time, the biodiversity regime's approach to intellectual property protection has evolved beyond the text of the CBD. The Conference of the Parties (COP) - the convocation of CBD members that determines how the Convention should be applied and implemented - has given detailed attention to harmonizing IPRs with the CBD's objectives. In particular, developing countries active in the COP, together with the support of nongovernmental organizations, have expressed concern about the adverse effects of IPRs and have sought to harness intellectual property rules to promote compliance with the Convention.

In response to this concern, official COP statements have stressed the need to "promot[e] increased mutual supportiveness and integration of biological diversity concerns and the protection of intellectual property rights." (CBD Decision IV/15, para. 9) Members of the COP have worked toward this objective by gathering information, commissioning case studies, holding workshops and drafting guidelines and recommendations. For example, in April 2002, the COP adopted the "Bonn Guidelines on Access to Genetic Resources and Fair and Equitable Sharing of the Benefits Arising Out of their Utilization." (CBD Decision VI/24, App. II, Annex, Part C) The Guidelines' most important recommendation encourages applicants for IPRs such as patents or plant breeders rights to disclose the country of origin of the genetic resources or the traditional knowledge upon which those IPRs are based. The Guidelines advocate these disclosures to monitor whether applicants for IPRs have obtained the prior informed consent of the country of origin and complied with the conditions of access (if any) that that country has adopted. (Id.)

Although the COP members adopted the Bonn Guidelines by consensus, the Guidelines' reliance upon intellectual property laws to promote compliance with the obligations of the CBD remains controversial. For example, a 2003 Communication from the European Commission to the European Parliament on the EC's implementation of the Bonn Guidelines considers whether to adopt "a self-standing disclosure requirement" for patent and plant variety protection applicants. (EC Implementation of Bonn Guidelines, COM (2003) 821) If adopted, such a disclosure requirement would not, however, be incorporated into national or European intellectual property laws. To the contrary, the failure to comply with the new disclosure requirement would, according to the EC Communication, "only have consequences outside the field of patent law." However, the EC Communication also states that the EC and its member states "should be ready to discuss, in the relevant international fora, the possibility of introducing *under intellectual property law* the same disclosure requirement... as a formal condition for patentability and not only as a self-standing obligation."

The Consultative Group on International Agricultural Research ("CGIAR"): CGIAR is an informal association of public and private donors founded in 1971 that supports an international network of agricultural research centres, each with its own governing body. CGIAR's mission is to engage in research, in partnership with other public and private entities, to promote sustainable agriculture in developing nations. With respect to plant genetic resources, CGIAR's principal method for achieving this objective is *ex situ* conversation. Networks of gene banks within CGIAR store and conserve seeds and propagating materials outside of their natural habitat for future use by farmers, researchers and breeders. The CGIAR network holds the world's largest *ex situ* collection of plant genetic resources for food and agriculture.

International Plant Genetic Resources Institute ("IPGRI") : IPGRI is the world's largest international institute dedicated to the conservation and use of plant genetic resources. Founded in 1974, IPGRI focuses on conservation, management and preservation of the diversity of plant genetic resources, through domestic, regional and international programmes and research initiatives.

The International Undertaking on Plant Genetic Resources (the "Undertaking") : FAO has helped to generate several nonbinding international instruments relating to plant genetic resources. The Undertaking, the first of these instruments, was adopted in 1983. As of 2000, 113 states were signatories to the Undertaking, thus pledging themselves to implement the recommendations it contains. For many years, the Undertaking served as the central legal instrument in FAO's global system for plant genetic resources, a system that includes a fund for the equitable sharing of benefits and a mechanism to give early warning about genetic resources under threat.

The Undertaking's principal objectives are to ensure that the need for conservation is globally recognized and that sufficient funds for this purpose are made available; to assist farmers and farming communities in the protection and conservation of PGRs and of the natural biosphere; and to allow farmers, their communities and countries to participate fully in the benefits derived from improved uses of PGRs, including through plant breeding. (WT/CTE/125, para. 11)

In its initial formulation, the Undertaking challenged a private property rights approach to plant genetic resources by declaring that all such resources, whether as cultivated by farmers in the field or modified through breeder innovations, were part of the "common heritage of mankind and consequently should be available without restriction." (art. 1). An interpretation issued adopted by FAO in 1989, however, clarified that plant breeders' rights were not incompatible with the Undertaking. It also recognized the interrelationship between the rights of traditional farmers (whose practice of saving seeds provided the raw genetic materials for innovation) and the rights of plant breeders (who use technology to achieve that innovation).

The International Treaty on Plant Genetic Resources for Food and Agriculture ("ITPGR") : On 3 November 2001, an intergovernmental conference sponsored by FAO adopted the text of a legally binding international agreement on plant genetic resources. The ITPGR entered into force on 29 June 2004. As of July of that year, the treaty had been ratified by 55 states and signed by an additional 50 countries. The ITPGR not only codifies and updates the nonbinding principles set out in the Undertaking and its subsequent revisions, but also contains provisions relevant to IPRs in plant genetic resources and plant varieties. Because of its importance to the subject of this study, an extended discussion of the ITPGR appears in section 4.3 below.

Specific Policy Objectives in Tension with IPRs

Having identified the principal institutions and agreements relating to plant genetic resources, the next sections address specific critiques of IPRs as applied to plants and plant varieties and the policy arguments that inform those critiques.

Preserving genetic diversity : Granting IPRs to plant breeders has uncertain consequences for preserving plant genetic diversity. Given the incentives that IPRs create for private parties to invest in research and breeding techniques relating to new plant varieties, it might be thought that IPRs would lead to an increase in plant genetic diversity over time.

A number of commentators have argued, however, that diversity is eroded rather than enhanced by granting IPRs to plant breeders. According to this view, *in situ* conservation by indigenous farmers diminished as they began to rely on commercial plant breeders for seeds and other propagating material. Rather than using informal breeding techniques to experiment with the creation of new varieties suitable for local growing conditions, indigenous farmers came to depend upon third party plant breeders to provide them with seeds possessing uniform genetic characteristics. The plant varieties that have come to dominate agriculture as a result of this dependence may possess many beneficial characteristics, but they do not enjoy the adaptive abilities of less well known and informally bred varieties.

To evaluate these competing assertions about the effects of IPRs on plant genetic diversity, empirical testing would be useful. Unfortunately, empirical studies of the causal impact of IPRs on plant genetic diversity are difficult to devise because of the uncertainty of screening out other relevant causal variables. Empirical studies that do exist often focus on industrialized countries rather than developing states and tend to rely on anecdotal evidence. As a result, the broader applicability of their conclusions is uncertain and often controversial.

Farmers' rights : A second challenge to IPRs concerns the relationship between farmers' rights and IPRs in plant varieties. The concept of farmers' rights was developed to reflect the contributions that traditional farmers, particularly in the developing world, have made to the preservation and improvement of plant genetic resources. FAO Resolution 5/89 defines farmers' rights as "rights arising from the past, present and future contributions of farmers in conserving, improving and making available plant genetic resources, particularly those in centres of origin/diversity." Such rights are also recognized in article 9 of the ITPGR.

Farmers' rights are in tension with IPRs for plant breeders because many farmers and farming communities do not claim exclusive rights in the cultivated landraces and plant varieties they have cultivated over time. Moreover, the subject matter requirements of existing IPR laws, which as explained below are designed to protect innovations in new and clearly distinguishable plant *varieties*, often cannot accommodate contributions of individual farmers using more informal methods to select for better crops or sought-after plant *characteristics*.

Advocates of farmers' rights have developed different approaches to address this situation and to reward farmers for their contributions to plant genetic diversity. The first approach involves situating the traditional practices of farmers as exceptions to the exclusive rights of plant breeders under existing IPR laws. In other words, breeders are precluded from demanding payment from farmers who engage in certain farming practices, such as saving seeds and planting seeds saved from prior purchases, or informally exchanging seeds. A second approach seeks to modify existing IPR laws so as to permit farmers themselves to claim exclusive rights in the plant varieties they cultivate informally. A third approach involves recognizing farmers' rights not through IPRs but through benefit sharing mechanisms, such as financial payments and technology transfers, that compensate farmers for their contributions to plant genetic diversity. This last approach - which is recognized in article 9 of the ITPGR- questions whether farmers in fact have "rights" as that term is understood within an intellectual property paradigm, while acknowledging the need to reward their contributions to plant genetic diversity.

Traditional knowledge: An issue closely related to farmers' rights is the recognition and protection of the plant-related knowledge, innovations and practices of indigenous and local communities. Advocates assert that those claiming IPRs in plant genetic resources and plant varieties often utilize such knowledge without adequately acknowledging the contributions of the communities who possess it. Mechanisms to redress this problem are similar to those discussed above relating to farmers' rights, a linkage reflected in article 9 of the ITPGR. They include modifying existing IPR laws to recognize traditional knowledge as a form of intellectual property and protecting traditional knowledge through *sui generis* national laws that recognize the communal ownership of much of traditional knowledge. In the fall of 2000, the World Intellectual Property Organization (WIPO) established a new Intergovernmental Committee on Intellectual Property and Genetic Resources, Traditional Knowledge and Folklore to study these issues.

Regulating access to plant genetic resources : Plant breeders and others seeking to develop plant-related innovations need access to existing stocks of plant germplasm for breeding, research and development. Access issues arise for both *in situ* and *ex situ* collections of seeds and plant propagating material.

With respect to *in situ* collections, a number of states have enacted national access laws and regulations to control third party access. These access controls are consistent with the CBD, which grants states the sovereign right to control their plant genetic resources. They also further the CBD's benefit sharing objectives by conditioning access upon the current or future payment of compensation or the transfer of technology to the state providing access. The specific conditions of access are often specified in Material Transfer Agreements ("MTAs") entered into between the entity (whether governmental or private) that owns or controls the germplasm and the entity (usually private) that seeks access to it.

Similar access issues arise with regard to *ex situ* collections of plant germplasm, including the seed banks maintained under the auspices of CGIAR. Although pursuant to a 1994 agreement between CGIAR and FAO the autonomous agricultural research centres under CGIAR's auspices each create their own access regulations, certain designated germplasm is to be held by the centres "in trust for the benefit of the international community." (art. 3(a)) With respect to

IPRs, the 1994 agreement provides that a centre "shall not claim legal ownership over the designed germplasm, nor shall it seek any [IPRs] over that germplasm or related information." (art. 3(b)). The centres also commit to apply the same restrictions on IPRs in MTAs with third parties to whom the centre transfers designated germplasm. (art. 10)

"Biopiracy" and property rights in unimproved plant materials : Issues of access to plant genetic resources are linked to larger controversies concerning the propriety of granting IPRs in the raw materials accessed.

Under settled principles of intellectual property law, unimproved plant germplasm already in the public domain cannot be removed and privatized. Nevertheless, there are several reported cases and one empirical study which indicate that plant breeders and commercial entities have on occasion been granted IPRs in wild plant varieties or in germplasm found in CGIAR seed collections. It is unclear whether such rights were granted because of inadvertence, insufficient or inaccurate information provided by the IPR applicant or particularities in national laws that limit the sources to which granting authorities may refer in determining what materials are in the public domain. Where such rights have been granted, however, states and interested NGOs have often succeeded in petitioning the authorities in other nations to remove legal protections they had previously granted.

Attempts to claim IPRs in unimproved plant genetic resources have often been labelled as a form of "biopiracy." Biopiracy is not a legal term of art, however and has been loosely used to refer to any act by which a commercial entity seeks to obtain IPRs over biological resources, including plant varieties, that are seen as "belonging" to developing states or indigenous communities. Thus, even where an IPR claim relates to improvements to raw plant materials, certain governments and NGOs have labelled the entity seeking legal protection as a biopirate if it has not provided a fair return to those who granted access to the raw materials.

Plant breeders' research interests : Even as between groups of plant breeders, the scope of IPRs in plant varieties can be controversial. Tensions arise between first generation breeders who have secured legal protection for new varieties and second generation breeders who seek to utilize those new varieties to develop still more varieties. As with farmers' rights, it is possible to use the exceptions and limitations provisions of national IPR laws to permit second generation innovators to engage in such activities without the authorization of first generation breeders.

International IPR Agreements Regulating Plant Varieties and Plant Breeders' Rights

This section identifies and discusses the provisions of three different international IPR agreements that protect plant varieties and plant breeders' rights.

The two major treaty systems that regulate these issues are the agreements established under the auspices of the *Union internationale pour la protection des obtentions vegetales* ("UPOV"), and the TRIPs Agreement included within the family of treaties administered by the World Trade Organization ("WTO"). These two treaty systems each contain a comprehensive set of rules for their members regarding IPRs over plant varieties.

In short, the UPOV treaties adopt a *sui generis* system of protection (that is, a system that is unique, or of its own kind) especially tailored to the needs of plant breeders. The TRIPs Agreement requires WTO Members to protect new plant varieties using patent rights, a *sui generis* system or some combination thereof. Because TRIPs provides states with this flexibility and because the treaty has an uncertain relationship to the previously adopted UPOV conventions, national governments face a wide array of options in choosing the intellectual property regime applicable to plant varieties. This section of the study outlines the requirements imposed by these two treaty systems, and Part III of the study then identifies and analyses these options.

The UPOV Acts

The first UPOV Act was drafted in 1961, principally by industrialized governments seeking to provide protections for plant breeders in their own and overseas markets. The UPOV was later revised in Acts adopted in 1972, 1978 and 1991.

As of July 2004, 55 states were parties to UPOV - 29 to the 1978 Act, 24 to the 1991 Act and 2 to the 1972 Act. (States Parties to the UPOV, Status on 30 June 2004). As a result, this study will focus on the two most recent UPOV Acts. Many of the accessions to these Acts are quite recent, with a number of developing states and countries in transition to a market economy acceding in the 1990s. The 1991 Act entered into force on 24 April 1998 and on that same date the

1978 Act was closed to future accessions except by a few states already in the process of adhering to it.

As explained in section 1.3.1 above, countries generally give domestic effect to the UPOV Act to which they are a party in one of two ways. In "automatic incorporation" states, courts and administrative agencies directly apply and enforce the Act, although implementing legislation is often needed to authorize administrative agencies to process applications to protect new plant varieties. In "legislative incorporation" states, by contrast, the UPOV Act does not become enforceable in domestic law until the state enacts a national plant variety protection law that conforms to the Act's requirements.

The 1978 UPOV Act

The 1978 UPOV Act adopts most of the international IPR obligations set out in Part I above, including a definition of the applicable subject matter and protected material, eligibility requirements, exclusive rights, national treatment, reciprocity, terms of protection and exceptions and limitations to exclusive rights. It does not, however, contain any provisions on MFN treatment or enforcement.

Subject Matter Requirements

Limited number of protected genera or species : Not all plant varieties must be protected under the 1978 Act. Rather, article 4 provides that member states are to progressively extend protection to an increasing number of genera or species, beginning with five on the date the treaty enters into force for that state and ending with twenty-four within eight years. In addition, member states are free to limit the Act's application within a particular genus or species to varieties with a particular manner of reproduction or multiplication, or a certain end-use.

Preclusion of dual protection with breeder's right and patent : The 1978 Act permits its signatories to protect plant varieties either with a distinct breeder's right or with a patent. However, article 2(1) precludes member states from granting both forms of protection "for one and the same botanical genus or species."

Protection of discovered varieties required : Although the 1978 Act focuses on plant varieties created through classical breeding methods, it is generally accepted that the treaty requires member states to protect varieties which have been discovered. This has been inferred from article 6.1(a) which indicates that a protected variety may result from a natural source of initial variation.

Eligibility Requirements

Assuming that a plant variety falls within a protected genera or species, it is eligible for protection under the 1978 Act only if it is: (1) new, (2) distinct from existing or commonly known varieties, (3) homogenous or uniform and (4) stable. (art. 6) When a variety fulfils these four criteria, it is listed in a national register or catalogue which publicly discloses that the variety is protected. (FIS/ASSINSEL, 2001a).

Novelty : To avoid protection for plant varieties that have already been exploited or are a matter of common knowledge, a new variety in which a breeder seeks protection must not have been sold on the market for more than a specified period of years prior to the date of application for protection. The 1978 Act specifies the maximum number of years during which such pre-application sales have occurred, with different periods of time set for different types of plants as well as for sales within the territory of the applicant state versus the territory of other states. (art. 6(1)(b)).

Distinctness : The 1978 Act states that a protectable plant variety must be "clearly distinguishable in one or more important characteristics from any other variety whose existence is a matter of common knowledge at the time when protection is applied for." (art. 6(1)(a)).

Although the treaty itself does not further define distinctness, the Guidelines for the Conduct of Tests for Distinctness, Homogeneity and Stability (UPOV Guidelines) use both qualitative and quantitative plant characteristics, including such visible attributes as leaf shape, stem length and colour, to determine if the difference between varieties is "clear and consistent." As explained below in the analysis of the UPOV 1991 Act, the concept of distinctness is critical to determining the scope of a breeder's right in plants that are closely related but not identical to a protected variety.

Homogeneity : Under the 1978 UPOV Act, a variety has to be "sufficiently homogeneous, having regard to the particular features of its sexual reproduction or vegetative propagation." (art. 6(1)(c)). The UPOV Guidelines further clarify that to be considered homogeneous, the variation shown by a variety must be "as limited as necessary to permit accurate description and assessment of distinctness and to ensure stability." The homogeneity requirement has been criticized by commentators as discouraging variability in plant varieties that are often useful for sound agricultural practices and as denying protection to breeders of cultivated landraces that exhibit diversity traits.

It is thus one of the sources for the criticisms identified in paragraph 1.3.6.1 above that plant breeders' rights are reducing plant genetic diversity by rewarding breeders of uniform plant varieties.

Stability : The stability requirement is a temporal one, requiring the breeder to show that the essential characteristics of its variety are homogeneous or uniform over time, even after repeated reproduction or propagation. (art. 6(1)(d)) In practice, what has been shown to be homogeneous is usually considered to be stable as well. For this reason, the stability requirement has engendered the same sort of critiques as the uniformity requirement in its preclusion of protection for cultivated landraces and other traditional plant varieties.

Protected material : The 1978 Act requires its signatories to protect a variety's reproductive or vegetative propagating material. The Act does not require protection of harvested material, with the exception of ornamental plants that are used for commercial propagating purposes. (art. 5(1))

Breeders' exclusive rights **:** Under article 5 of the 1978 Act, any person seeking to engage in the following three acts, with respect to a protected variety's reproductive or vegetative propagating material, must obtain the prior authorization of the breeder: (1) production for purposes of commercial marketing, (2) the offering for sale and (3) marketing. The 1978 Act does not, however, require member states to extend these exclusive rights to harvested material or other marketed products.

National treatment and reciprocity : Member states must grant the three exclusive rights referred to in the preceding paragraph in the same manner to both national breeders and to breeders who reside in or are nationals of other 1978 Act member states. However, where a state extends legal protection to a specific genus or species, or where it provides more extensive exclusive rights to breeders than the rights required under the treaty, reciprocity is permitted. Thus, a state providing these additional rights may restrict protection to breeders from those member states that apply the Act to the same genus or species, or that provide such additional exclusive rights to their own nationals. (arts. 3 and 5(4))

Term of protection : The 1978 Act requires a minimum term of protection of fifteen years, with the exception of vines, forest trees, fruit trees and ornamental trees, which are protected for no less than eighteen years.

Exceptions and limitations : Two major exceptions and limitations to exclusive rights exist under the 1978 Act: (1) a breeders' exemption and (2) a farmers' privilege. The Act also permits members to impose compulsory licences.

Breeders' exemption: This exemption in article 5(3) precludes member states from granting to breeders of protected varieties the right to authorize or refrain from authorizing other breeders seeking to use the protected variety to create new varieties or to market those new varieties. States are permitted to grant breeders such an authorization right only if the repeated use of the protected variety is necessary for the commercial production of the new variety. According to the International Association of Plant Breeders and the International Seed Federation, this breeders' exemption "is essential for continued progress from plant breeding." (FIS/ASSINSEL, 1999)

Farmers' privilege : The focus of the 1978 Act on commercial exploitation of protected plant varieties has been interpreted to allow the use of seeds and propagating material for noncommercial purposes without the breeder's prior authorization. In national plant variety protection laws, this implicit noncommercial exception most frequently benefits farmers who purchase the seeds of protected varieties. The scope of this so-called farmers' privilege varies widely, however. Some nations only permit farmers to plant seeds saved from prior purchases to be used on their own land holdings, while others allow them not only to replant but also to sell limited quantities of seeds for reproductive purposes, a practice often referred to as "brown bagging."

Compulsory licences in the public interest : Article 9 of the 1978 Act permits members to restrict breeders' exclusive rights for "reasons of public interest." Where such restrictions are enacted to ensure the widespread distribution of the variety (such as where the breeder fails to supply the demand for the variety in a reasonable quantity and price or unreasonably refuses to license the variety to third parties), the breeder must receive equitable remuneration.

The 1991 UPOV Act

The limited scope of the 1978 Act led a number of member states of the UPOV to adopt a revised Act with enhanced rights for plant breeders. The major revisions of the 1991 Act are discussed below:

Subject Matter Requirements

Phased-in protection of all genera or species : Unlike its predecessor, the 1991 Act requires states to protect at least fifteen plant genera or species upon ratifying or acceding to the Act, and to extend protection to all plant varieties within ten years. (art. 3(2)) It also contains a definition of a plant "variety" as a "plant grouping within a single botanical taxon of the lowest known rank" which can be "defined by the expression of the characteristics resulting from a given genotype or combination of genotypes; distinguished from any other plant grouping by the expression of at least one of the said characteristics; and considered as a unit with regard to its suitability for being propagated unchanged." (art. 1(vi)) No definition of "variety" appears in the 1978 Act, which indicates that member states to that earlier treaty have greater discretion in defining the characteristics of plant groupings that qualify for protection.

Dual protection with breeders' right and patent permitted : In response to demands from breeders in industrialized counties, the 1991 Act removed the 1978 Act's ban on dual protection and now permits member states to protect the same plant variety with both a breeders' right and a patent.

Protection of discovered varieties : The 1991 Act makes explicit the 1978 Act's implicit requirement that discovered varieties be protected. It does so through article 1(iv)'s definition of a "breeder" as including a "person who bred, or discovered and developed, a variety."

Eligibility Requirements

The four eligibility requirements that must be demonstrated to merit protection for a specific variety - novelty, distinctness, uniformity and stability - are preserved in the 1991 Act, subject to only minor changes in scope and wording. (arts. 7-9) Thus, the 1991 Act has received the same criticism as the 1978 Act for its encouraging of genetic standardization and its inability to protect more diverse plant varieties, traditional varieties or cultivated landraces.

Breeders' Exclusive Rights in Protected Material

Extensive additions to the 1978 Act were made with respect to the exclusive rights enjoyed by breeders in protected material of plant varieties.

Enumerated exclusive rights in propagating material : The breeder's prior authorization must be obtained for the use of reproductive or vegetative propagating material of the variety for (1) production or reproduction, (2) conditioning for the purpose of propagation, (3) offering for sale, (4) selling or marketing, (5) exporting, (6) importing and (7) stocking for any of these purposes.

Extension of rights to harvested material and products : These exclusive rights apply not only to propagating material but also to harvested material, where the harvest has been obtained through an unauthorized use of the propagating material and the breeder has not had a reasonable opportunity to exercise his or her right in relation to that material. In addition, member states are permitted but not required to provide additional rights to breeders, including rights in products made directly from harvested material.

National Treatment Required

Whatever the particular exclusive rights member states adopt in their plant variety protection laws, those rights must also be provided to the nationals of other 1991 Act member states. Unlike the 1978 Act, granting rights only on condition of reciprocity is not permitted. (art. 4) For example, if a 1991 Act member chooses to grant more expansive exclusive rights to breeders than the exclusive rights required by the 1991 Act, it must grant those rights to breeders from all other 1991 Act member states. The state cannot choose to grant such expansive rights to only those 1991 Act states that also provide the same level of rights protection to the first state's nationals.

Terms of Protection

The 1991 Act extends the term of protection to 20 years, and requires a 25-year term for tree and vine varieties.

Exceptions and Limitations

A second area of major revision concerns the reduced scope of exceptions and limitations to breeders' exclusive rights, which is found in article 15 of the 1991 Act.

Private, noncommercial exception : The 1991 Act makes explicit what was only implicit in the 1978 Act, namely that private, noncommercial activities with respect to new varieties are outside of the breeder's control. This exception would presumably permit subsistence farmers to use protected seeds and other propagating material for their own consumption.

Research exception : The 1991 Act also recognizes that breeders cannot restrict "acts done for experimental purposes." This exception would permit research and testing of protected varieties for scientific purposes that does not lead to commercial exploitation.

Limited breeders' exemption : Like its predecessor, the 1991 Act recognizes the right of breeders to use protected varieties to create new varieties. However, this exception is itself restricted in its application to such new varieties as are not "essentially derived" from protected varieties. (arts. 14(5) and 15) The drafters added this restriction to prevent second generation breeders from making merely cosmetic changes to existing varieties in order to claim protection for a new variety.

The concept of essential derivation has proved highly controversial in practice, however. Breeders have been unable to agree on a definition of the minimum genetic distance required for second generation varieties to be treated as not essentially derived from an earlier variety and thus outside of the first breeder's control. (FIS/ASSINSEL, 1999). Nevertheless, the overall effect of this provision of the 1991 Act has been to narrow the breeders' exemption and expand the IPRs of first generation breeders.

Limited farmers' privilege : The 1991 Act recognizes an explicit farmers' privilege, but one that is limited in scope. In particular, each member state may enact a provision in its national plant variety protection laws permitting farmers to use for propagating purposes "on their own holdings" the product of the harvest which they obtained by planting a protected variety "on their own holdings." In addition, this privilege must be exercised "within reasonable limits and subject to the safeguarding of the legitimate interests of the breeder." (art. 15(2)) Several points can be inferred from this language.

First, unlike the 1978 Act, the 1991 version of the farmers' privilege does not authorize farmers to sell or exchange seeds with other farmers for propagating purposes, a limitation that commentators have criticized as inconsistent with the practices of farmers in many developing nations, where seeds are exchanged for purposes of crop and variety rotation.

Second, according to the International Association of Plant Breeders, the "reasonable limits" language in article 15(2) requires states to restrict the acreage, quantity of seed and species subject to the farmers' privilege, while the requirement to safeguard breeders' "legitimate interests" requires farmers to pay some form of remuneration to the breeder for their privileged acts. (FIS/ASSINSEL, 2001a) The latter assertion is controversial, however, and has not been enacted in the national laws of all the 1991 Act member states. For example, the EC's Council Regulation on Community Plant Variety Rights, No. 2100/94 of 27 July 1994, requires such payments except in the case of small farmers, whereas the United States' Plant Varieties Protection Act of 1994, 7 U.S.C. Sec. 2543, does not.

Compulsory licences in the public interest : Article 17 of the 1991 Act contains a compulsory licence provision similar to that found in the 1978 Act. It permits members to restrict breeders' exclusive rights only for reasons of public interest and requires payment of equitable remuneration to the breeder whose rights are limited.

Resistance to the 1991 Act

Many developing nations, particularly those in Africa, have resisted ratifying the 1991 Act or adopting it as the standard for their plant variety protection laws. The foreign ministers of the more than 50-member Organization for African Unity (now the African Union) issued a statement at a January 1999 meeting calling for a hold on IPR protection for plant varieties until an Africa-wide system has been developed that grants greater recognition to the cultivation practices of indigenous communities.

However, at a subsequent meeting of the Organisation Africaine de la propriete Intellectuelle (OAPIO), patent officials from sixteen francophone African nations recommended that their countries adopt the 1991 Act. (Machipisa, 1999). As of the date of this study, Tunisia, Kenya and South Africa are the only African UPOV member states. Tunisia ratified the 1991 Act in August 2003, whereas the other two countries are parties to the 1978 Act. Other developing states parties to the 1978 Act are Bolivia, China, Ecuador, Kyrgyzstan, Mexico, Nicaragua, Panama, Paraguay, Trinidad and Tobago and Uruguay.

These and other developing countries are, however, under pressure to ratify the 1991 Act as part of bilateral or regional trade and investment agreements.

The TRIPs Agreement

Although the UPOV Acts have provided IPR protection for plant varieties for more than forty years, their significance has recently been overshadowed by a different intellectual property treaty, the Agreement on Trade-Related Aspects of Intellectual Property Rights ("TRIPs" or the "TRIPs Agreement"). Adopted in 1994 as a treaty administered by the WTO, TRIPs is the first and only IPR treaty that seeks to establish universal, minimum standards of protection across the major fields of intellectual property, including patents, copyrights, trademarks, industrial designs, integrated circuits and trade secrets. Although the TRIPs Agreement devotes only minimal attention to plant breeders' rights or plant variety protection and does not even mention the UPOV Acts, its adoption has done more to encourage the legal protection of plant varieties than any other international agreement.

TRIPs as a Spur to Plant Variety Protection

TRIPs' influence on plant variety protection stems from the following sources: (1) its link to other international trade agreements; (2) its widespread adherence by states in both the industrialized and developing world; (3) its novel enforcement, review and dispute settlement provisions; (4) the requirement in TRIPs article 27.3(b) that its signatories must provide protection for plant varieties "either by patents or by an effective *sui generis* system or by any combination thereof"; and (5) a formal review of article 27.3(b) which was scheduled to be held in 1999. The following paragraphs briefly address each of these five issues. Subsequent sections devote more detailed treatment to the protection of plant varieties with patents and to the elements necessary to create an "effective *sui generis* system."

The Relationship of TRIPs to the WTO and the International Trading System

Unlike all prior intellectual property treaties, TRIPs is not a free-standing agreement concerned solely with IPRs. Rather, TRIPs is linked to a larger family of trade-related agreements concerning subjects such as trade in goods and services, agriculture, textiles and health-related restrictions on imports. All of these agreements were adopted within the WTO during the Uruguay Round of trade negotiations held between 1988 and 1994. As such, TRIPs was part of a global "package deal." Industrialized nations secured a commitment from developing nations to provide minimum standards of effective legal protection to intellectual property products, and in exchange developing nations received a commitment from industrialized countries to open their domestic markets to goods and other products manufactured in the developing world.

Widespread Adherence to TRIPs

The result of this global bargain was widespread adherence to all WTO Agreements, including the TRIPs Agreement. As of July 2004, 147 states or customs territories were obligated to comply with TRIPs by virtue of their membership in the WTO. In addition, many states (particularly those in the developing world) who became bound by TRIPs had not ratified earlier intellectual property treaties and were thus required to make significant changes to their national laws to bring them into compliance with the Agreement. In acknowledgement of this fact, TRIPs contains a phase-in provision pursuant to which most of the treaty's substantive obligations became binding for developed nations in 1996, but did not become binding on developing nations and nations in transition to market economies until 2000. Least developed nations are given the most leeway and are not required to implement the treaty's substantive obligations until 2006.

The Enforcement, Review and Dispute Settlement Provisions of TRIPs

In addition to its widespread adherence, the influence of the TRIPs Agreement can be traced to its unique provisions relating to the enforcement of IPRs within national laws, the review of those national laws by the TRIPs Council and the mechanism for settlement of disputes between states leading to rulings backed up by the threat of trade sanctions.

Enforcement provisions : Unlike prior IPR agreements, TRIPs does not only specify the minimum substantive requirements for various forms of intellectual property. In addition, it requires WTO Members to adopt "effective" provisions within their national laws to permit the owners of intellectual property products to enforce their rights against those who infringe them. (art. 41.1) These enforcement provisions include detailed judicial and administrative remedies, border measures and criminal procedures. (articles 41 to 61) To take just one example applied to plant varieties, a breeder whose new variety is sold commercially without permission must be able to bring a civil judicial action seeking an injunction to stop the conduct of the unauthorized seller and to recover damages from him or her.

Review provisions : Article 68 establishes a Council for TRIPs to "monitor the operation of th[e] Agreement and... Members' compliance with their obligations" under the treaty. Since its founding, one of the Council's principal functions has been to formally review the national laws, regulations and judicial decisions of WTO Members in each area of intellectual property covered by TRIPs.

Several goals are served by the TRIPs Council's review functions. First, the reviews create an incentive for governments to bring their national laws and practices into compliance with the Agreement so that they may present positive information to the Council. Second, the reviews identify for both the reviewed state and for other WTO Members areas of the law which may not be in full compliance with the treaty. And finally, TRIPs Council reviews provide a critical opportunity to publicize national laws and practices relating to IPRs which might otherwise be difficult to obtain.

In the area of plant varieties, the TRIPs Council has gathered and organized a considerable amount of useful information concerning national government practices. In December 1998 the Council prepared a detailed list of questions concerning plant variety protection. It asked WTO Members who were already obligated to protect plant varieties to respond to these questions by indicating the manner in which their national laws provide such protection, with other WTO Members requested to use their best efforts to furnish such information. The Council later received and published responses from 17 states and from the European Communities and its member states. (WTO Job No. 2689 of 7 May 1999). In 2001 and 2002, the Council reissued these questions and invited WTO Members that had not already done so to provide information on their plant variety protection practices. Additional responses were received from six states. Information distilled from some of these responses is discussed in section 3.3 below.

Dispute settlement provisions : Another significant innovation of TRIPs is its dispute settlement system. Although prior intellectual property agreements, including the UPOV Acts, contained provisions for filing of complaints against non-complying treaty parties before the International Court of Justice, no state ever invoked this dispute settlement option. It was widely believed that pursuing such a complaint would be perceived as an unfriendly act, would be time-consuming and costly and that states were unlikely to implement decisions of the court. The TRIPs Agreement removes each of these concerns by linking to a WTO Agreement known as the Dispute Settlement Understanding which contains a streamlined dispute settlement mechanism that is one of the most widely used and effective interstate dispute resolution systems in international law. Indeed, the mere existence of such a system creates strong incentives for WTO Members to bring their national laws into compliance with the WTO Agreements, including TRIPs, to avoid the possibility of dispute settlement proceedings.

These strong incentives notwithstanding, some countries may not comply. In such a case, where one WTO Member believes that another Member has failed to fulfill its obligations under the Agreement, it may initiate consultations under the Dispute Settlement Understanding with a view to resolving the dispute. If the parties fail to reach agreement, the complaining state may then request the WTO to convene a three-member panel of experts to review its allegations. Such panels generally issue a decision within six months. The panel's decision is adopted by the WTO's Dispute Settlement Body unless the losing party or parties elect to appeal the decision to the seven-member Appellate Body, a standing tribunal of seven trade experts who are authorized to review the panel's findings and issue a decision within three months. If either the Appellate Body or an unreviewed panel decision concludes that a WTO Member has violated the TRIPs Agreement, the Dispute Settlement Body will recommend that the defending state bring its national laws into compliance with the treaty. If the state fails to do so, the complaining state may then commence an arbitral proceeding to specify the amount of compensation that the defending state must pay to remedy the violation, or in lieu of such compensation the complaining state may seek authorization to impose trade sanctions on the non-complying Member.

As of July 2004, WTO Members (mostly developed nations) had commenced 24 dispute settlement proceedings against both developed and developing nations concerning a wide variety of intellectual property issues. (WTO Dispute Settlement: Index of Disputes Issues: 15 July 2004) Approximately one third of these disputes were resolved prior to a decision by a panel, with the defending state agreeing to modify its laws to bring them into compliance with the Agreement. In other cases, however, WTO Members complied only after a panel or the Appellate Body had issued a decision against them. And in a few instances, Members that have not modified their laws instead have negotiated settlements involving the payment of compensation to the complaining Member.

As of the date of this study, there have been no TRIPs dispute settlement proceedings relating to intellectual property protection for plant varieties. Whether such proceedings will be brought in the future depends on a number of variables, many of which are still uncertain.

Three interrelated reasons suggest that a dispute settlement proceeding concerning plant variety protection is a likely possibility: first, the significant difference of views among WTO Members over the scope of legal protection to be provided to plant varieties; second, the fact that developing country WTO Members first became obligated to protect plant varieties only in 2000; and third, the absence of plant variety protection in many developing countries' national laws prior to the 2000 deadline. Taken together, these variables suggest that the plant breeders of a developed WTO Member who learn that their protected varieties are being exploited without authorization in developing WTO Members due to inadequate national laws will pressure their governments to file a complaint against such states to compel them to adhere to their treaty obligations.

Other factors, however, suggest that a complaint over plant varieties is unlikely to be filed, particularly within the next five to ten years. First, the initiation of the Doha Round of trade negotiations in November 2001 (discussed in Part IV below) is likely to cause WTO Members to show restraint in their dispute settlement strategies while the multi-year negotiations are proceeding. This is especially so given that the Doha Round will likely provide WTO Members with the first meaningful opportunity to harmonize other international obligations (such as those set forth in the CBD and the ITPGR) with the plant variety and other IPR protection requirements of the TRIPs Agreement. Second, past practice has shown that WTO Members are most likely to file dispute settlement complaints where they are pressured to do so by the owners of intellectual property products. It is unclear whether plant breeders can make a sufficient showing that their overseas markets are being harmed by other states' failure to provide adequate legal protection for plant varieties. Third, WTO Members are generally reticent about filing complaints where their probability of success is unclear. Therefore, given the uncertain scope of protection for plant varieties currently required by TRIPs, if Members enact *some form* of plant variety protection in their national laws, other Members are unlikely to challenge those laws unless they can identify a clear violation of the Agreement. Nevertheless, the potential for dispute settlement proceedings (and the trade sanctions they can engender) will continue to provide strong incentives for states to enact plant variety protection laws.

The Unique Scope of Plant Variety Protection Required by TRIPs

Article 27.3(b) contains the only textual provisions of the TRIPs Agreement relating to plant variety protection. The article states in relevant part:

Members may also exclude from patentability: (b) *plants* and animals other than microorganisms; and essentially biological processes for the production of plants or animals other than non-biological and microbiological processes. However, *Members shall provide for the protection for plant varieties either by patents or by an effective* sui generis *system or by any combination thereof.* (emphasis added)

The meaning of this article has been the subject of significant debate among both WTO Members and nongovernmental organizations (NGOs) with differing views over the propriety of IPR protection for plant varieties. A detailed review of that debate and an analysis of article 27.3(b) is provided below. For present purposes, two overarching points are worth noting.

First, TRIPs' provisions on plant varieties do not refer to or incorporate any preexisting intellectual property agreements, including the 1978 and 1991 UPOV Acts. This omission contrasts sharply with other fields of intellectual property, such as patents, copyrights and trademarks, for which TRIPs expressly requires WTO Members to comply with the standards of protection contained in preexisting IPR agreements, such as the Berne Convention for the Protection of Literary and Artistic Works and the Paris Convention for the Protection of Industrial Property. As a result of this omission, WTO Members are neither required to become members of UPOV nor to enact national laws consistent with either UPOV Act in order to comply with their obligations under TRIPs. Although the drafting history of TRIPs does not explain this markedly different treatment of plant varieties, it seems likely that compliance with UPOV was not required because so few WTO Members were party to UPOV and those who were could not agree upon which of its two most recent Acts should serve as the standard for protection.

Second, article 27.3(b) permits WTO Members to protect plant varieties using one of three distinct approaches: (1) patent law, (2) an effective *sui generis* system or (3) a combination of elements from both systems. Thus, unlike most other areas of intellectual property protected by TRIPs, article 27.3(b) expressly grants Members significant discretion to choose the manner in which they will protect plant varieties and it contemplates that that discretion may be exercised differently by different states.

This discretion and the opportunity for divergent outcomes it engenders has significant consequences. On the one hand, TRIPs' failure to incorporate and build upon the preexisting UPOV Acts may have "a deharmonizing effect," (Correa 1994a) with states within the UPOV system enacting one type of plant variety protection law and states outside of that system enacting a different kind of law (which may or may not resemble each other). This could create significant complexities and uncertainties for plant breeders seeking to market protected varieties in different jurisdictions. On the other hand, this sanctioned diversity of legal approaches allows WTO Members to balance the protection of plant breeders' rights against the other important and competing societal goals identified in Part I, many of which are found in other international agreements. Seen from this perspective, article 27.3(b) provides a much needed "safe space" for governments to harmonize conflicting norms and policies - a space that is lacking in other areas of the TRIPs Agreement.

The Review of TRIPs Article 27.3(b)

The last sentence of article 27.3(b) states that "[t]he provisions of this subparagraph shall be reviewed four years after the date of entry into force of the WTO Agreement." Inasmuch as the WTO Agreement entered into force on 1 January 1995, the review contemplated by this article should have been conducted in 1999. The review was commenced but was not concluded, largely as a result of disputes between industrialized and developing nations over the scope of the review process. Governments submitted additional information and proposals to the TRIPs Council in 2000 and 2001, but no formal action was taken. With the launching of the Doha Round of trade talks in November 2001, however, the review of article 27.3(b) recommenced in earnest.

In June 2002, eleven developing states submitted a detailed proposal to amend TRIPs to "prevent systematic conflicts with the CBD arising from the implementation of TRIPS." (WTO Doc. IP/C/W/356, para. 11) The proposal seeks to compel all WTO Members to require applicants for patents relating to biological materials and traditional knowledge to disclose certain information as a condition of obtaining legal protection. (Id., para. 10)

The European Communities and Switzerland responded to the developing states' proposal with a compromise. The EU response calls for the negotiation of "a self-standing disclosure requirement," which, while not functioning as a new eligibility criterion for patent protection, "would allow Members to keep track, at [the] global level, of all patent applications with regard to genetic resources for which they have granted access." In June 2003, developing countries rejected this compromise and "reaffirm[ed] and strengthen[ed] their demand for a strong disclosure of origin mechanism" within TRIPs that would require "not only detailed information about who provided the [genetic] materials or the [traditional] knowledge used [in patent applications], but also positive proof of benefit sharing and of prior informed consent." African countries also proposed a new "Decision on Traditional Knowledge" to be incorporated into TRIPs.

"TRIPs Plus" Bilateral and Regional Agreements

In the decade since TRIPs was first adopted, the United States and the European Community have negotiated a growing number of bilateral and regional trade and investment treaties with developing countries. Many of these treaties contain provisions concerning IPRs. Commentators have referred to these treaties by the appellation "TRIPs plus" because they (1) contain intellectual property protection standards more stringent than those found in TRIPs, (2) obligate developing countries to implement TRIPs before the end of its specified transition periods or (3) require such countries to accede to or conform to the requirements of other multilateral intellectual property agreements. As one study recently noted, "[t]o the extent that ["TRIPs plus" treaties] include these additional aspects, they are pushing harmonization forward at a pace that is greater than is apparently possible within the framework of the WTO." (OECD, 2003) In addition, once a developing state enacts this higher level of IPR protection in its national laws, TRIPs' MFN clause obligates that state to extend the same protection to *all other WTO Members*. In this way, bilateral agreements can be used to ratchet up the level of protection for IPRs beyond what is required in the TRIPs Agreement.

Several "TRIPs plus" treaties contain provisions relating to plant variety protection that go beyond the minimum level of intellectual property protection specified in TRIPs. For example, the recently concluded U.S.-Central America Free Trade Agreement (applicable to Costa Rica, El Salvador, Guatemala, Honduras, Nicaragua and possibly to the Dominican Republic), the 2002 U.S.-Chile free trade agreement, the 2000 U.S.-Jordan free trade agreement and the 2000 EU-Mexico free trade agreement, as well as certain Euro-Mediterranean Association agreements and the draft Free Trade Agreement of the Americas, all mandate UPOV as the appropriate mechanism to protect plant breeders' rights. (GRAIN, 2004b) The treaties also require these countries to ratify the 1991 UPOV Act within specified time frames. Such obligations limit the discretion enjoyed by these countries to tailor their national laws to achieve policy objectives in tension with IPRs.

Patent Protection for Plant Varieties Under TRIPs

The TRIPs Agreement mandates its signatories to provide patent protection for any inventions in all fields of technology, provided that the inventions are "new, involve an inventive step and are capable of industrial application." (art. 27(1)) However, with regard to plant-related inventions, TRIPs permits Members to exclude from patentability altogether "plants," "essentially biological processes for the production of plants" and "plant varieties." (art. 27.3(b)) Thus, as presently written, the TRIPs Agreement would permit WTO Members to decline to protect plant varieties with a patent (provided, as discussed in section 2.3.3 below, that they protect such varieties with an effective *sui generis* right).

Applicability of Patent Protection for Plant Varieties

For this reason, it may initially appear that governments can ignore patents as they consider how to protect plant-related innovations in their national legal systems. Nevertheless, a basic understanding of the patent provisions of the TRIPs Agreement and a comparison of those provisions with the plant breeders' rights approach contained in the UPOV Acts is essential for a number of reasons.

First, extending patent protection to plant-related inventions and innovations remains an option for national governments. Recall that TRIPs' "minimum standards" framework expressly contemplates that WTO Members may provide greater protection for IPRs than are mandated by the Agreement. Article 27.3(b) in particular invites Members to protect plant varieties with patents or with a combination of patents and a *sui generis* system.

A number of governments in the industrialized world, including the United States, Japan, Australia, New Zealand, Sweden and the United Kingdom, have capitalized on this opportunity by permitting plant breeders to obtain patent protection in new varieties provided that the eligibility requirements for a patent have been met.

Second, because TRIPs does not require any patent protection for plant-related innovations, it follows as a matter of course that the treaty does not compel WTO Members to adopt any particular form of patent protection. This allows governments the option of including plant varieties within their existing utility patent statutes and/or of enacting a separate statute applicable exclusively to plants. (Utility patents are generally granted to "any new and useful process, machine, manufacture or composition of matter, or any new and useful improvement thereof... " In making this decision, however, it will be useful for governments to understand the traditional elements of patent law and the ways in which they may be varied within the confines of the TRIPs Agreement. The United States, for example, has enacted multiple plant variety protection laws.

It was one of the first nations to adopt, in the Plant Patent Act of 1930, a unique form of protection applicable to asexually reproducing plants. In addition, since the mid 1980s, U.S. courts and the U.S. Patent and Trademark Office have concluded that plant breeders may also seek standard utility patents under the U.S. Patent Act. In December 2001, the U.S. Supreme Court confirmed that breeders could apply for both forms of protection with respect to the same variety. Such a result is prohibited by the 1978 UPOV Act but permitted under the 1991 Act to which the U.S. is a party.

Third, an understanding of patent principles is also necessary because of the difficulty in some nations of distinguishing between plant varieties (which are often excluded from national or regional patent laws) and other plant-related innovations (to which patent protection may be extended).

In Europe, for example, article 53(b) of the European Patent Convention prohibits the patenting of "plant varieties" as such. Nevertheless, the European Patent Office recently confirmed that claims to patent protection that are broadly drawn to encompass "plants" or an invention broader than a single variety *may* be patented, even though such claims may encompass multiple varieties. As a result, it may be possible for plant breeders in Europe to fashion their claims to receive *de facto* patent protection for new plant varieties.

There are significant differences in approach between plant breeders' rights and patents. In the case of plant breeders' rights, the eligibility requirements for protection are not onerous, but the scope of protection granted is quite narrow, both in terms of exclusive rights and the various exceptions and limitations to those rights. Patent laws strike a very different balance.

Eligibility requirements are high and difficult to meet, but once granted a patent conveys broad rights to exclude third parties from exploiting the patented invention. Depending on the needs and level of development of plant breeder industries within its territory, a government may decide that either or both forms of protection will provide the appropriate incentives to encourage plant-related research and innovation.

Fifth, and finally, a review of the TRIPs Agreement's plant-related patent rules will soon occur in the context of the Doha Round of trade negotiations. Such a review may produce an agreement to amend TRIPs to require WTO Members to extend patent protection to plant-related innovations or plant varieties. National governments which have thus far eschewed patent protection would then need to revise their national laws to make such protection available, in addition to or instead of the legal protection provided under national plant variety protection laws.

At a minimum, therefore, governments interested in promoting plant-related innovations should be apprised of the relevant legal issues so that they may contribute in an informed way to the upcoming discussions.

Subject Matter Eligibility Requirements

According to the text of the TRIPs Agreement and authoritative commentary interpreting it, plant-related innovations may be patented if they are "inventions" that are novel, non-obvious and useful. Although commentators frequently assert that these cumulative requirements impose significant barriers to the patenting of plant varieties, in practice breeders have succeeded in patenting not only new varieties, but also hybrid plants and inbred and hybrid plant lines. As two commentators recently noted, since 1985 the U.S. Patent and Trademark Office "has granted hundreds of utility patents on all aspects of innovation relating to plant science: plants themselves, seeds, breeding methods and plant biotechnology."

Inventions : The TRIPs Agreement does not define the term "invention," thus leaving its precise meaning to national patent laws or regional patent rules such as those operating within the European Union.

All such laws recognize that an invention must be more than a mere discovery of a natural phenomenon or naturally occurring substance. Thus, for example, a breeder who merely identified the existence of a previously unknown wild variety or a cultivated landrace that was known only to an isolated indigenous community should not be permitted to patent that variety or landrace.

The trend among the industrialized countries, however, and in the United States, Japan and the countries of the European Union in particular, is to recognize that an isolated and purified form of a natural substance may be patented. For example, article 3.2 of the 1998 Directive on the Legal Protection of Biotechnological Inventions states that "biological material which is isolated from its natural environment or produced by means of a technical process" may be patentable. (Directive 98/44/EC of 6 July 1998, O.J. L213) Perhaps because this approach involves making a very fine distinction between discoveries and inventions, many developing countries have declined to follow it, instead choosing to deny patent protection to plant materials found in nature even if they have been isolated or purified by human intervention. According to one commentary, this patent ban is compatible with the TRIPs Agreement provided that it does not extend to plants with modified or artificial plant gene sequences, which often significantly vary from naturally occurring substances and thus are properly classified as inventions.

A second attribute of a patentable invention is that an applicant seeking such a patent must disclose the invention in a sufficiently clear and complete manner that permits a person skilled in the art to carry out the invention. (TRIPs, art. 29(1)). In addition to ensuring that the claimed invention in fact meets patent eligibility requirements, disclosure also permits third parties to access the invention for purposes of improving it or developing new inventions. In the case of plant-related innovations, national laws permit inventors to satisfy the disclosure requirement either by a written description of the invention and/or by a deposit of the protected material (i.e. seeds, germplasm or other biological material). TRIPs does not require national governments to adopt any particular form of disclosure, nor does it specify the timing, manner or conditions of third party access. As a result, national laws vary widely on this point.

Novelty, inventive step and industrial application : These three cumulative eligibility requirements are mandated by article 27(1) of the TRIPs Agreement and are found in all national patent laws. Even among states with strong patent protection laws, however, the interpretation and application of each of these steps varies considerably.

Novelty : The principal objective of the novelty requirement is to ensure that the claimed invention cannot be found in the "state of the art" or "prior art" already in existence. These terms refer to the body of knowledge which is available to the public before the date of filing of an application for patent protection. Patent examiners establish both the novelty and inventive step requirements by comparing the claimed invention to the state of the art or prior art. A critical issue for plant-related innovations concerns the form in which prior art exists and the difficulty of accessing traditional knowledge as prior art.

Under United States patent law, for example, novelty can be negated if "the invention was patented or described in a printed publication in this or a foreign country or in public use or on sale in this country... " (35 U.S.C. Sec. 102(b)) The important point is description: foreign *uses* of claimed inventions are not considered as part of prior art. The Patent Cooperation Treaty (PCT) adopts a similar approach, limiting prior art to "everything which has been made available to the public anywhere in the world by means of written disclosure." (Rule 33.1 of Regulations under the PCT) Thus, in the United States and in countries following the PCT, even where foreign plant-related innovations are known, used or disclosed in other than written form, it may be possible for the inventor to obtain patent protection.

Article 54(2) of the European Patent Convention adopts a different approach, defining the state of the art to encompass everything made available to the public by means of written or oral descriptions, by use or indeed in any other way, prior to the filing of the application. Nevertheless, the absence of systemic databases concerning traditional knowledge, including knowledge relating to plants, has led examiners in various countries to grant patents for inventions that were later revoked after they were discovered to encompass prior art.

Inventive step : The inventive step requirement looks to the state of the art to determine whether the claimed invention is not obvious to a person skilled in the art. Although the majority of international treaties and national laws defined inventive step in this way, the application of the standard to plant-related innovations varies and may be a significant hurdle to patent protection given "the enormous speed of technological progress" in the field.

Industrial application : The industrial application requirement is concerned with an invention's practical utility, and in particular whether "it can be made or used in any kind of industry, including agriculture." (E.g. European Patent Convention, art. 7) This eligibility requirement does not appear to pose a significant obstacle to the patenting of new plant varieties, given their ready use in the fields of plant breeding, horticulture and agriculture.

Exclusions from Patentability

As noted above, article 27.3(b) of TRIPs permits WTO Members to deny all patent protection to "plants" and "plant varieties." Nevertheless, a state that chooses to provide such protection may nevertheless choose to exclude from patentability a narrower category of inventions, "the prevention within their territory of the commercial exploitation of which is necessary to protect *ordre public* or morality, including to protect... plant life or health or to avoid serious prejudice to the environment." (Art. 27(2)) Some commentators have asserted that this language grants states the discretion to exclude broad categories of inventions that further the erosion of genetic diversity on the theory that issues of public policy and morality include a state's sovereign right to control its natural resources.

However, because WTO dispute settlement panels have strictly interpreted clauses in the WTO Agreements that permit states to derogate from protected treaty rights, the more persuasive view is that exclusions on this basis must be fact-specific, narrowly drawn and no broader than required to achieve their purpose. (Moufang, 1998) Resolution of this debate in the context of plant-related innovations will not occur until the WTO dispute settlement bodies issue a ruling on the matter, an unlikely prospect so long as Members are free to exclude plants and plant varieties entirely from patentability.

Exclusive Rights

Once a patent is awarded, the owner of the patented product or process enjoys a broad panoply of exclusive rights. These include the right to prevent third parties from making the product, using the process or from using, offering for sale, selling or importing for those purposes the patented product or the product obtained by the patented process. (TRIPs, art. 28) States that choose to extend patent protection to plant-related innovations would thus be expected to protect products composed of plants and parts of plants and processes (including biological processes) for the production of plants. Although there is some dispute over whether traditional plant breeding methods (as opposed to creating varieties through genetic manipulation) can be protected by process patents, the practice of states that do extend patent protection to plants reveals that such methods have been protected.

Term of Protection

TRIPs article 33 imposes a minimum 20-year term of protection from the date the patent application is filed.

Exceptions and Limitations

As compared to limitations on plant breeders' rights permitted under the UPOV, the limitations on a patent owner's exclusive rights permitted under the TRIPs Agreement are far narrower. These limitations can be divided into exceptions to exclusive rights and compulsory licenses, which permit certain uses by third parties but require remuneration to the patent owner.

Exceptions : Article 30 of TRIPs permits WTO Members to adopt "limited exceptions" to exclusive patent rights provided that they do not "unreasonably conflict with a normal exploitation of the patent" and "do not unreasonably prejudice the legitimate interests of the patent owner, taking account of the legitimate interests of third parties." This three-part test is applicable to all patents protected by the TRIPs Agreement.

Canadian Patents Case : A WTO dispute settlement panel clarified the meaning of article 30 in Canada - Patent Protection of Pharmaceutical Products (Generic Medicines), WT/DS114/R (17 March 2000). The panel considered a complaint brought by the European Communities to two provisions of Canadian patent law. The first permitted uses of patented pharmaceuticals without the patentee's authorization for the purpose of obtaining approval of a generic product before the patent's term expired.

The second permitted production and stockpiling of the generic product for release immediately after the expiration of the patent. Both procedures were designed to permit the marketing of generic versions of pharmaceuticals promptly after the expiry of the patent. The panel decision confirmed that only the first exception is consistent with TRIPs. More importantly, the panel concluded that any exception that results in "a substantial curtailment" of the patent owner's exclusive rights is inconsistent with article 30. (WT/DS114/R, para. 7.36)

If TRIPs were amended to require patent protection for new plant varieties, the standard adopted by the Canadian Patents panel would preclude Members from enacting many of the exceptions that are permitted under plant variety protection laws, such as the breeders' exemption and the farmers' privilege. The same prohibition on these privileges already exists in several countries that currently protect plant varieties and other plant-related innovations in their national patent laws.

Research and experimentation exception : Most national patent laws permit third parties to engage in experimentation or research related to the patented invention. Some states, however, including the United States, narrowly construe this exception to preclude unauthorized commercialization of any products or processes that result from such research if they encompass the patentable invention. In such countries, commercial activities permitted under the breeders' exemption found in plant variety protection laws are prohibited where the variety is protected by a patent.

Thus, for example, a breeder would not be permitted to cross patented seeds to produce improved varieties. Even in patent systems in which the experimental use exception tolerates some commercialization (Cornish, 1998), such as those found in Europe, it is likely that the commercial acts allowed under a traditional breeders' exemption would "conflict with a normal exploitation of the patent" and would thus amount "a substantial curtailment" of his or her exclusive rights in violation of TRIPs article 30. Thus, if patent protection of plant varieties were to become mandatory under TRIPs, a breeders' exemption such as that found in article 22,V of Mexico's patent law would be unlikely to survive a challenge before a WTO dispute settlement panel.

Farmers' privilege : A similar fate is likely to befall the exception permitting farmers to save and reuse seeds on their own land without the patent owner's permission, another exception traditionally allowed under plant variety protection laws. The United States does not recognize a farmers' privilege under its utility patent laws. Although such an exception has been adopted in article 11 of the European Union's 1998 Biotechology Directive (conditioned upon the payment of equitable remuneration except by small farmers), at least one commentator has questioned whether that exception is compatible with TRIPs article 30, because such an exception would unreasonably prejudice the "legitimate interests" of the patent owner.

Compulsory licences : TRIPs contains a complex set of rules that regulate when states may compel patent owners to license their products and processes to governments or to private parties. Although the TRIPs Agreement does not specify the grounds which justify the creation of compulsory licences, because article 5A(2) of the Paris Convention is incorporated by reference into article 2(1) of TRIPs, it can be inferred that such licences may be granted only to prevent "abuses which might result from the exercise of the [patent owner's] exclusive rights."

Even where such abuses exist, TRIPs article 31 imposes further conditions upon the granting of compulsory licences, including individual consideration of each case, prior negotiations with the patent owner seeking a voluntary licence, limitations on the scope and duration of compulsory licences and requiring their termination when the circumstances leading to their creation are no longer in effect. Most importantly, patent owners must receive "adequate remuneration," taking into account the value of the rights licensed. One area in which compulsory licences may affect plant breeders is that of dependent patents, which are defined as patents whose use requires the authorization of an earlier patent owner. Such patents are prevalent in plant breeding, where the creation of new varieties often occurs incrementally in the form of adaptations and improvements of existing varieties, as opposed to radically new innovations. Because incremental innovation often requires access to protected varieties, governments may seek to impose compulsory licences in favour of third party breeders who are unable to negotiate voluntary access to patented plant varieties. However, the compatibility of such licences with TRIPs is untested and governments seeking to impose them should ensure that they comply scrupulously with each of the many requirements set forth in article 31.

Because the patent compulsory licence provisions of TRIPs are far more detailed and narrow than the compulsory licences permitted under the 1978 and 1991 UPOV Acts, commentators are divided over their usefulness in the area of plant-related innovations. Some argue that such licences will be largely irrelevant to the field, whereas others claim that governments may adopt such licences "to ensure access to patented materials in order to attain specific agricultural objectives (e.g. availability of a given material for farmers) or food security...." Even if licences to achieve these objectives are permitted, however, the restrictions that TRIPs imposes significantly limit the ability of WTO Members to grant such licences in favour of third parties.

Sui Generis Protection for Plant Varieties under TRIPs

As noted above, TRIPs authorizes WTO Members to eschew patent protection for plants and plant varieties and adopt instead an "effective *sui generis* system" of protection. The following discussion identifies the core requirements of such a system, laying the groundwork for Part III which outlines the permissible options available to WTO Members seeking to grant IPRs to plant varieties while simultaneously seeking to achieve other objectives.

Identifying and Evaluating Competing Proposals for *Sui Generis* Systems

TRIPs does not define the term *sui generis* as it appears in article 27.3(b), nor does the treaty's drafting history shed any light on its meaning. As a general matter, the term *sui generis* is understood to mean "of its own kind" or "unique," a definition which does little to identify what sort of unique legal systems are permitted under the treaty. This lack of definitional clarity has allowed advocacy groups and commentators to advance different and often competing proposals for how effective *sui generis* protection should be structured, producing significant confusion for governments seeking to understand and implement their obligations under the TRIPs Agreement.

The discussion below identifies and evaluates the competing proposals. In brief, it concludes that if WTO Members choose to protect plant varieties using a *sui generis* right as opposed to a patent, they are required to create a distinct IPR applicable to plant varieties that complies with the core requirements and objectives of the TRIPs Agreement.

Sui generis systems that do not contain these core elements or conflict with them are not TRIPs-compatible and expose the WTO Members who adopt them to challenges under the WTO dispute settlement system. Nevertheless, the mandatory requirements of article 27.3(b) preserve significant leeway for national governments to work out the precise manner in which they will balance protection of IPRs against other international obligations and national objectives.

Sui Generis Systems as Defined by the UPOV Acts

Several commentators assert that a state adopting national legislation in compliance with either the 1978 UPOV Act or the 1991 UPOV Act has satisfied its obligations under article 27.3(b). Those advocating for compliance with the 1991 Act stress the more extensive IPR protections it provides to plant breeders, while commentators endorsing the 1978 Act as the appropriate *sui generis* standard stress that only the earlier Act was in force at the time TRIPs was being negotiated.

In fact, however, the protection required by the two UPOV Acts is neither necessary nor sufficient for an "effective *sui generis* system." It is not necessary because article 27.3(b) does not require plant variety protection laws to contain the same subject matter, eligibility requirements, exclusive rights, terms of protection or other detailed provisions of either of the two UPOV Acts. And it is not sufficient because TRIPs requires WTO Members to structure their national IPR laws in ways that the two UPOV Acts do not.

Note, however, that developing counties that are parties to a "TRIPs plus" bilateral or regional trade or investment treaty may be required, by virtue of those treaties, either to join UPOV or to implement the standards set forth in one of the UPOV Acts. A country that has undertaken such obligations does *not* possess the same flexibility in designing a *sui generis* system to protect plant varieties as a country which has not undertaken such obligations.

For example, as explained in paragraphs 2.2.2.4 and 2.2.3.4 above, both the 1991 Act and the 1978 Act adopt the principle of national treatment, the same principle required by TRIPs. But under the UPOV Acts such treatment need be extended only to the nationals and residents of *other UPOV member states* and to legal persons having their headquarters in such states. Any WTO Member that limited national treatment in this way would unquestionably violate article 3 of TRIPs, which requires extension of national treatment to all other WTO Members whether or not they are also UPOV members.

However, to comply with article 3, a state whose plant variety protection law contains a UPOV-specific national treatment rule can easily amend its law to expand the number of states to which national treatment applies. Other UPOV-sanctioned rules may require more extensive changes.

For example, the 1978 Act allows a state to impose a reciprocity requirement in certain circumstances, limiting the right to apply for protection of a new variety to nationals or residents of those member states which also apply the Act to the genus or species to which that variety belongs. This reciprocal protection is incompatible with TRIPs' national treatment rule, and possibly its most favoured nation principle as well, both of which are inconsistent with a system of reciprocity.

That TRIPs' drafters did not intend either UPOV Act to be the exclusive model for *sui generis* protection of plant varieties is confirmed by their failure to refer to the Acts anywhere in the Agreement. By contrast, where the drafters intended Members to comply with standards found in preexisting international IPR treaties, they stated so expressly. Nevertheless, as explained in sections 3.3.1 and 3.3.2 below, most provisions of the two UPOV Acts are fully consistent with an effective *sui generis* system, and most countries have in fact adopted plant variety protection laws that are consistent with one or both UPOV Acts.

Core Requirements of an Effective Sui Generis System

According to a 1997 report prepared by two leading commentators, there are four core elements that any national plant variety protection law must contain in order to qualify as an "effective *sui generis* system" within the meaning of TRIPs article 27.3(b):(1) the law must apply to all plant varieties in all species and botanical genera; (2) it must grant plant breeders an IPR, i.e. the exclusive right to control particular acts with respect to those protected varieties, or at a minimum, the right to remuneration when third parties engage in certain acts; (3) it must provide national treatment and MFN treatment to breeders from other WTO Members; and (4) it must contain procedures that enable breeders to enforce the rights granted to them under such a law. The paragraphs that follow discuss each of these four requirements.

Protection of all plant varieties : Although both UPOV Acts require protection of only a specific number of plant varieties (paras. 2.2.1.1 and 2.2.3.1 above), a careful reading of the TRIPs Agreement reveals that WTO Members are required to protect all varieties. article 27.3(b) states that "Members shall provide for the protection of *plant varieties*" (emphasis added) without any further qualification or limitation. Inasmuch as article 27.3(b) is included in a provision of the TRIPs Agreement which carves out numerous and specific exceptions to patent protection, if the drafters had intended to allow states to limit the number or type of plant varieties subject to protection, they would have said so expressly.

Indeed, the drafters were aware that both UPOV Acts permitted precisely such limitations, yet they chose neither to refer to those Acts nor to their limited scope of protection. As explained in paragraph 3.4.2.1 below, however, the obligation to protect all varieties does not require states to provide *the same level of protection* to each and every variety.

Protection of plant varieties with an IPR : If protection granted to plant varieties must take the form of an IPR, WTO Members would enjoy virtually unbounded discretion to choose the manner in which to protect plant varieties, for example through a system of taxation on seeds. The WTO's Appellate Body recently interpreted the text of the TRIPs Agreement, however, to require protection through an IPR. Nevertheless, because some commentators and NGOs have questioned whether plant variety protection is in fact a form of intellectual property included in the TRIPs Agreement, a brief discussion of this point is needed.

Article 1(2) of TRIPs defines "intellectual property" for purposes of the Agreement as "all categories of intellectual property that are the subject of Section 1 through 7 of Part II" of TRIPs. Protection for plant varieties appears in Section 5 relating to patents, and is thus a form of intellectual property protected by the Agreement. This interpretation is confirmed by the United States - Section 211 Omnibus Appropriations Act of 1998, AB-2001-7, WT/DS176/AB/R (2 January 2002). In that case, a WTO panel concluded that trade names were not a category of intellectual property protected by TRIPs because they were not expressly referred to in article 1(2) (an omission that also applies to plant varieties). The Appellate Body reversed this finding. As an example, it referred to Members' option to protect plant varieties "by *sui generis* rights (such as breeder's rights) instead of through patents," and concluded that *sui generis* rights were in fact a form of intellectual property protected by the treaty. (Id., para. 335)

The significance of this conclusion is that IPRs for plant varieties must take the form of other IPRs protected by the TRIPs Agreement. Specifically, governments must either grant to the owners of protected varieties (1) the right to exclude all third parties from engaging in certain activities with respect to those varieties (an exclusive rights approach) or, (2) at a minimum, the right to receive equitable remuneration when a third party engages in such activities.

Nothing in TRIPs requires states to adopt the exclusive rights approach, nor does it indicate which exclusive rights must be granted to new varieties. Nevertheless, the vast majority of IPRs are protected by exclusive rights, with compulsory licences used only in limited areas such as new industries or distribution networks, abuses of monopoly powers or where high transaction costs preclude private licensing mechanisms from developing. In addition, a number of economic studies have suggested that granting exclusive rights to the owners of intellectual property products is a more efficient mechanism for creating incentives to innovate than a compulsory licence approach, in which political factors or government inertia are likely to produce skewed incentives. (Merges, 1996)

***National and MFN treatment applicable to rights granted* :** At least one respected commentator has argued that national and MFN treatment are inapplicable to a *sui generis* system protecting plant varieties. The theory advanced is that these two forms of treatment are inapplicable to IPRs not covered by the TRIPs Agreement and that plant variety protection is an IPR that is not covered by TRIPs. However, the text of TRIPs article 3(1), as interpreted by WTO dispute settlement jurists, indicates that plant variety protection must adhere to such obligations.

Thus, in the United States - Section 211 Omnibus Appropriations Act of 1998, AB-2001-7, 2 January 2002, the WTO Appellate Body stressed the critical importance of the national and MFN treatment rules in multilateral trade agreements (paras. 242 and 297), and it concluded that the obligation to grant such treatment applies to all subjects of intellectual property protected by TRIPs, a designation that includes *sui generis* protection of plant varieties. As a result, in the area of plant variety protection, each Member must grant "no less favourable treatment" to the nationals of all other WTO Members than it grants to its own nationals and it must also grant to such foreign nationals "any advantage, favour, privilege or immunity" provided to any other country. (TRIPs, arts. 3.1 and 4) Applied to plant varieties, this would mean that any IPR that a state grants to its own plant breeders must be granted to breeders from all other WTO Members, and that any IPR granted to breeders from one WTO Member must be granted to breeders from all WTO Members.

Enforcement of rights by private parties : The only qualification that article 27.3(b) imposes upon *sui generis* systems for protecting plant varieties is that they be "effective." Although TRIPs does not define this term, it refers to it in Part III dealing with enforcement of other IPRs against acts of infringement by third parties. For this reason, a WTO dispute settlement panel is unlikely to find a *sui generis* system to be effective unless it provides a meaningful opportunity for private parties to enforce their rights in protected varieties.

Specific enforcement measures are not mandated by the Agreement, however. Had the drafters intended greater specificity, they could have easily made a reference in article 27.3(b) to the detailed enforcement rules applicable to other IPRs in Part III of TRIPs. Although a *sui generis* system must be "effective," this does not mean that a state must devote a particular level of resources to enforcement measures. In particular, article 41.5 of TRIPs makes clear that the treaty does not create "any obligation with respect to the distribution of resources as between enforcement of intellectual property rights and the enforcement of law in general." As a result, what qualifies as "effective" in a state with abundant resources and well-developed systems of law enforcement will not be required of states with more limited resources or weaker law enforcement systems.

Additional requirements of an "effective"* sui generis *system? : Beyond effective enforcement measures, it is uncertain whether a WTO dispute settlement panel would find that article 27.3(b) imposes additional requirements for a *sui generis* system to qualify as "effective." In general, international tribunals have concluded that rights granted in a treaty must be interpreted so as to make them effective rather than illusory.

Thus, if a WTO Member were to grant very limited and weak rights to breeders together with broad and extensive exemptions in favour of farmers or other users of plant germplasm, its plant variety protection law might not survive WTO scrutiny. Such a possibility is likely to arise only in extreme cases in which a WTO Member has enacted a plant variety protection system that is functional in theory but non-functional in practice.

States that implement the core TRIPs requirements in good faith - that is, states that grant breeders intellectual property rights and enforcement measures applicable to varieties in all species and botanical genera and that provide those same rights and measures to breeders from other WTO Members - are unlikely to have their laws challenged successfully.

Options Available to National Governments under Existing International IPR Agreements Protecting Plant Varieties and Plant Breeders' Rights

In order to identify the options available to national governments under international IPR agreements, this section begins by categorizing states according to the agreements to which they are parties. This classification scheme takes into account the fact that a state may be bound by more than one IPR-related treaty, a fact that may significantly limit its discretion. The section then identifies both the mandatory actions required of national governments in each category as well as the discretionary choices available to them. It also identifies trends in national laws which illustrate the manner in which states are complying with their particular set of international obligations.

Classification of States According to their International IPR Obligations

The discretion enjoyed by states to shape their plant variety protection laws to balance the protection of IPRs against other societal concerns is dependent upon the international agreement or agreements to which they are parties. The following table identifies that discretion at a broad level of generality. It does so by listing the mandatory components of plant variety protection that a state must adopt depending upon whether it is a member of (1) WTO only; (2) WTO and the 1991 UPOV Act; (3) WTO and the 1978 UPOV Act; and (4) either the 1991 UPOV Act or the 1978 UPOV Act only; or (5) no IPR agreement relating to the protection of plant varieties.

Note that some developing states, by virtue of having ratified a "TRIPs plus" bilateral or regional trade or investment treaty, may be required to adhere to some or all of the provisions contained in either the 1991 UPOV Act or the 1978 UPOV Act, *even if they are not members of UPOV itself*. Developing states that are parties to such "TRIPs plus" treaties should carefully consult the text of these agreements to determine the precise scope of their supplementary IPR obligations, including for purposes of applying the classification scheme set forth below. For example, a developing country that is a WTO Member but not a member of UPOV may nevertheless be required to adhere to all of the provisions of the 1991 UPOV Act by virtue of having ratified a "TRIPs plus" treaty. In such a case, the state would be obligated to implement the mandatory components of plant variety protection contained not only in TRIPs but also in the 1991 UPOV Act, and its discretion to balance the protection of IPRs against other competing policy objectives would be reduced accordingly.

Cumulative Treaty Obligations

Thus, for example, a state that is a party to a UPOV Act and also to the TRIPs Agreement must comply with its obligations under both treaties. Given that each treaty contains somewhat different but mostly cumulative and consistent rules concerning the protection of plant varieties, a state that is a party to both treaties will necessary have more extensive obligations to protect plant varieties and thus will enjoy concomitantly less discretion.

Conflicting Treaty Obligations

Where the provisions of two treaties are in direct conflict, the rule is far less settled. If two agreements relate to the same subject matter and the states parties to both agreements are the same, article 30 of the Vienna Convention on the Law of Treaties specifies that the agreement that is later in time is given effect. The obligations in the TRIPs Agreement are thus likely to prevail over any conflicting obligations in the 1978 UPOV Act, such as the ban on protecting varieties within the same genus or species with both a breeders' right and a patent. (1978 Act, art. 2(1)) Because the 1991 Act entered into force in 1998 and TRIPs entered into force in 1995 it may be argued that the 1991 UPOV Act is the later-in-time agreement.

However, it is unlikely that any conflicts issue will arise between the two treaties, inasmuch as nothing in TRIPs requires conduct that the 1991 UPOV Act forbids. (Such a conflict might arise, however, in the unlikely event that TRIPs were interpreted to impose more restrictive exclusions to breeders' exclusive rights than the mandatory exclusions found in article 15 of the 1991 UPOV Act.) It can therefore be argued that the two treaty systems are fully compatible, with TRIPs merely augmenting the plant variety protection requirements of the UPOV Acts.

Identifying the Level of Discretion Available to States in each Treaty Classification

In general, the discretion a state enjoys in limiting the scope of intellectual property protection provided to plant varieties increases.

WTO and UPOV 1991 Act Members

States that are members of both TRIPs and the 1991 UPOV Act have the least discretion. This designation currently applies to 27 nations: Australia, Austria, Belarus, Bulgaria, Croatia, the Czech Republic, Denmark, Estonia, Finland, Germany, Hungary, Israel, Japan, Kyrgyzstan, Latvia, Lithuania, the Netherlands, Moldova, Poland, Romania, Singapore, Slovenia, South Korea, Sweden, Tunisia, the United Kingdom and the United States of America.

Mandatory Requirements

These 27 states must extend protection to all plant varieties, comply with TRIPs' national MFN treatment obligations and adopt effective enforcement measures. The adoption of these rights - (1) production or reproduction, (2) conditioning for the purpose of propagation, (3) offering for sale, (4) selling or marketing, (5) exporting, (6) importing and (7) stocking for any of these purposes - fully satisfies their obligation under TRIPs to protect varieties with an IPR. Finally, states that have agreed to these dual obligations must also comply with all of the other provisions of the 1991 Act, including its terms of protection and limitations on the breeders' exemption and farmers' privilege.

Options and Implementation Issues

A state whose plant variety protection laws are already in compliance with the 1991 Act needs to take only limited steps to comply with article 27.3(b) of TRIPs, including immediate protection of all varieties and extension of national treatment to all WTO Members. The most difficult hurdle is likely to involve enacting effective mechanisms for plant breeders to enforce their rights. However, far more detailed enforcement measures must be adopted for the other forms of intellectual property protected by TRIPs. A state may therefore choose to extend to breeders the same domestic enforcement opportunities that it is required to provide to the owners of other IPRs.

National Implementation

In information submitted to the TRIPs Council in May 1999, the following states indicated that they had adopted plant variety protection laws modelled on the 1991 Act: Australia, Bulgaria, Members of the European Communities, Japan, Morocco, Poland, Romania, South Korea, South Africa, Switzerland and the United States of America. (Other states parties to both UPOV 1991 and TRIPs did not submit information to the Council.) Of the states adopting such laws, Morocco, South Africa and Switzerland had not ratified the 1991 Act.

Bilateral IPR Agreements and UPOV Membership

A separate issue concerns whether WTO Members will continue to join UPOV after the entry into force of the TRIPs Agreement. Some commentators predicted that UPOV membership would stagnate as states capitalized on the greater discretion provided to them under the *sui generis* option of article 27.3(b). Since TRIPs entered into force in 1995, however, 28 nations have become parties to the 1978 or 1991 UPOV Acts. (States Parties to the UPOV, Status on 30 June 2004) The continued expansion of UPOV membership can be traced in part to the recent proliferation of "TRIPs plus" treaties.

WTO and UPOV 1978 Act Members

This designation currently applies to 24 nations: Argentina, Bolivia, Brazil, Canada, Chile, China, Colombia, Ecuador, France, Ireland,

Italy, Kenya, Mexico, New Zealand, Nicaragua, Norway, Panama, Paraguay, Portugal, Slovakia, South Africa, Switzerland, Trinidad and Tobago and Uruguay. States in this category enjoy somewhat greater discretion as a result of the more limited protection of plant breeders' rights contained in this earlier UPOV Act.

Mandatory Requirements

States parties to both agreements must extend protection to all plant varieties, comply with TRIPs' national and MFN treatment rules and adopt effective enforcement measures. They must also comply with all of the other 1978 Act requirements, including its eligibility requirements, terms of protection, exclusive rights and mandatory breeders' exemption. As compared to the 1991 Act, however, breeders' exclusive rights are more limited, terms of protection for varieties are shorter and exceptions and limitations are broader.

Options and Implementation Issues

States that become Members of the WTO after joining the 1978 UPOV Act and adopting laws to comply with that Act face a similar situation to states that are parties to both the WTO and the 1991 Act. To fully comply with TRIPs, these 1978 Act member states must modify their national laws to protect the four core requirements of article 27.3(b) and they must remove all provisions of their laws which impose a reciprocity requirement as a condition for protecting varieties of foreign breeders. In addition, states in this category may choose to modify their laws to incorporate some or all of the standards found in the 1991 Act without actually becoming a member of that Act. Their refusal to do so, however, does not violate article 27.3(b), inasmuch as the standards found in the 1978 Act satisfy their obligation to protect plant varieties with a *sui generis* IPR.

Trends in National Laws

In information submitted to the TRIPs Council in May 1999, Canada, the Czech Republic, Hungary, Norway and New Zealand (the only states submitting information) indicated that they had adopted plant variety protection laws that complied with the 1978 Act. At the time of these submissions, the Czech Republic and Hungary were members of the 1978 Act. They have since acceded to the 1991 Act.

1991 or 1978 UPOV Act Members Only

There are few states outside of the WTO that protect plant varieties solely by virtue of being UPOV members. And both are observer governments with the WTO. These nations face a somewhat different set of international legal obligations.

Mandatory Requirements

They must comply with all of the numerous requirements of the UPOV Act to which they are a party. However, they have no obligation to provide effective enforcement measures to breeders, their national treatment obligations are limited to those states that are members of the same UPOV Act and, in the case of those states bound by the 1978 Act, they need not protect all plant varieties and may impose certain reciprocity requirements on those varieties they do protect.

Options and Implementation Issues

Provided that their laws are in compliance with the relevant UPOV Act and unless they have ratified other international IPR agreements, these states need make no additional changes to their national plant variety protection laws.

Member of WTO Only

States that are parties to the TRIPs Agreement but are not members of either UPOV Act enjoy considerably greater discretion than those that are members of either UPOV Act alone. By virtue of their adherence to TRIPs, these states are required only to comply with the four core obligations of article 27.3(b). However, those countries that have ratified "TRIPs plus" treaties that require them to comply with one of the UPOV Acts). Given the large number of states in this category (and the additional number of states likely to enter this category if their efforts to join the WTO succeed), a more detailed discussion of the numerous options available to them in balancing protection of plant breeders' rights against the competing policy objectives is provided in section 3.4 below.

Not a Member of WTO, UPOV or other IPR Agreements

Nations in this category have no obligation to protect plant varieties or breeders' rights in any form. Nevertheless, the mandatory requirements and discretionary choices facing other governments are likely to become matters of significant concern to these states as the world trading and intellectual property protection systems expand. Attention to IPR issues will be of particular importance when these states implement the obligations of other international treaties to which they may be parties, such as the biodiversity goals of the CBD or the access requirements of the ITPGR. States in this category should consider whether it is sound policy to enact laws that are incompatible with TRIPs and the UPOV, since such an action will preclude their membership in these treaty systems unless they undertake a further revision of their national laws.

In addition, unilateral actions by or bilateral agreements with industrialized nations may pressure these states to comply with international IPR standards even if they remain outside of the WTO or UPOV. For these reasons, states that are not yet Members of the WTO should give careful attention to the discretionary choices available to WTO Members analysed in the next section.

Chapter 8

Trans-acting RNA

Trans-acting siRNA (ta-siRNA or tasiRNA) is a form of small interfering RNA (siRNA) that represses gene expression through post-transcriptional gene silencing in land plants. It is transcribed from the genome to form a polyadenylated, double-stranded precursor. Unlike other siRNAs, but like microRNA, ta-siRNA does not require full complementarity to its target RNA to have an effect.

RasiRNA

Repeat associated small interfering RNA (rasiRNA) is a class of small RNA that is involved in the RNA interference (RNAi) pathway. RasiRNA is a subclass of Piwi-interacting RNAs (piRNAs), which are small RNA molecule which interact with Piwi proteins. Piwi proteins are an Argonaute subfamily. RasiRNA associate with both the Ago and Piwi Argonaute protein subfamily while piRNA only associates with the Piwi Argonaute subfamily. In the germline, RasiRNA is involved in establishing and maintaining heterochromatin structure, controlling transcripts that emerge from repeat sequences, and silencing transposons and retrotransposons.

Classification

There are at least three Argonaute subfamilies that have been found in eukaryotes. Unlike the Ago subfamily which is present in animals, plants, and fission yeast, the Piwi subfamily has only been found in animals. RasiRNA has been observed in *Drosophila* and some unicellular eukaryotes but its presence in mammals has not been determined unlike piRNA which has been observed in many species of invertebrates and vertebrates including mammals; however, since proteins which associate with rasiRNA are found in both vertebrates and invertebrates, it is possible that active rasiRNA exist and have yet to be observed in other animals. RasiRNAs have been observed inSchizosaccharomyces pombe, a species of yeast, as well in some plants both of which have not been observed to contain the Piwi Argonaute protein subfamily. It has been observed that both rasiRNA and piRNA are maternally linked, but more specifically it is the Piwi protein subfamily that is maternally linked and therefore leads to the observation that rasiRNA and piRNA are maternally linked.

RasiRNA are Distinct from other RNAi Pathways

RasiRNA is distinct from other RNAi pathways such as microRNA (miRNA) and small interfering RNA (siRNA) as well as from piRNA. Unlike miRNA and siRNA which function through the Ago Argonaute protein subfamily, RasiRNA function through the Piwi Argonaute protein subfamily. RasiRNA is also distinct in its size. Contrary to miRNAs which are 21-23 nucleotides in length, siRNAs which are 20-25 nucleotides in length, and piRNAs which are 24-31 nucleotides in length, rasiRNAs are 24-29 nucleotides in length depending on the organism it came from.

Unlike siRNA which are derived from both the sense and antisense strand, rasiRNA are derived from the antisense. Interestingly, while miRNA requires Dicer-1 for its production, and siRNA requires Dicer-2, rasiRNA does not require either; however, in some plants there are Dicer-like (Dcl) proteins that have been identified where Dcl1 produces 24 nucleotide miRNA and siRNA while Dcl2 produces 24 nucleotide rasiRNA. This research shows that not only is rasiRNA production distinct from miRNA and siRNA, but that rasiRNA may be found in plants while Piwi proteins are not.

RasiRNA Biogenesis

It is presumed that the source of rasiRNA is double stranded RNA produced by annealing of sense and antisense related transposable elements. The biogenesis of rasiRNA is independent of Dicer, but does require the Argonaute proteins Argonaute 3 (Ago 3), Piwi and Aubergine which is a Piwi-like protein. The mechanism for rasiRNA biogenesis is a ping-pong mechanism. The Piwi/Aub associated RNA is the rasiRNA. The rasiRNAs match the antisense strand of retrotransposons and repetitive sequence elements (hence the name rasiRNA). The Ago3 associated RNAs are derived from the sense strand. The ping-pong mechanism which is observed in this image is the mechanism for the generation of the 5' end of rasiRNA while the generation of the 3' end of rasiRNA is still unknown.

Importance of RasiRNA

While miRNA act in translational repression and mRNA cleavage and siRNA act in mRNA cleavage, rasiRNA act to regulate chromatin structure and transcriptional silencing. In Drosophila, mutations in the Piwi proteins that associate with rasiRNA lead to sterility and loss of germline cells in both males and females. Transposon repression is not affected by the loss of Dicer within the germline cells revealing that this is the target of the rasiRNA pathway. Similar to miRNA and siRNA, the rasiRNA silencing pathway is evolutionarily conserved and homology dependent. When the rasiRNA pathway is not present, germline cells may undergo retrotransposition which are sensed as DNA damage and signal the cell to apoptosis. RasiRNA is key to the regulatory mechanism of many organisms as part of the RNA interference pathway.

History and Discovery

Small RNAs guiding RNA silencing pathways were first discovered in 1993 in Caenorhabditis elegans and since then have been observed in many organisms. RasiRNA were discovered sometime after 1993 and observed in both Drosophila melanogaster and Schizo-saccharomyces pombe and since have been observed in some single celled eukaryotes and some plants.

7SK RNA

7SK, also known as RN7SK, was discovered as an abundant small nuclear RNA gene found in metazoans in the mid-70s but a possible function has only recently been suggested.

Two independent studies have found that 7SK RNA binds the CDK9/cyclin T complex (known as elongation factor P-TEFb). P-TEFb activates transcription byphosphorylating the C-terminal domain of RNA polymerase II. This process is negatively regulated by the 7SK RNP.

The secondary structure of 7SK RNA shown here was proposed by Wasserman and Steitz on the basis of chemical modification and enzymatic cleavage experiments. This family contains a number of pseudogenes and 7SK derived repeats.

Retrotransposon

Retrotransposons (also called transposons via RNA intermediates) are genetic elements that can amplify themselves in a genome and are ubiquitous components of the DNA of many eukaryotic organisms. They are a subclass of transposon. They are particularly abundant in plants, where they are often a principal component of nuclear DNA. In maize, 49-78% of the genome is made up of retrotransposons.

In wheat, about 90% of the genome consists of repeated sequences and 68% of transposable elements. In mammals, almost half the genome (45% to 48%) comprises transposons or remnants of transposons. Around 42% of the human genome is made up of retrotransposons while DNA transposons account for about 2-3%.

Biological Activity

The retrotransposons' replicative mode of transposition through an RNA intermediate increases the copy numbers of elements rapidly and thereby can increase genome size. Like DNA transposable elements (class II transposons), retrotransposons can induce mutations byinserting near or within genes. Furthermore, retrotransposon-induced mutations are relatively stable, because the sequence at the insertion site is retained as they transpose via the replication mechanism.

Retrotransposons copy themselves to RNA and then back to DNA that may integrate back to the genome. The second step of forming DNA may be carried out by a reverse transcriptase which the retrotransposon encodes.

Transposition and survival of retrotransposons within the host genome are possibly regulated both by retrotransposon- and host-encoded factors, to avoid deleterious effects on host and retrotransposon as well, in a relationship that has existed for many millions of years between retrotransposons and their plant hosts. The understanding of how retrotransposons and their hosts' genomes have co-evolved mechanisms to regulate transposition, insertion specificities, and mutational outcomes in order to optimize each other's survival is still in its infancy.

Most retrotransposons are very old and through accumulated mutations, are no longer able to retrotranspose.

Types of Retrotransposons

Retrotransposons, also known as class I transposable elements, consist of two sub-types, the long terminal repeat (LTR) and the non-LTR retrotransposons.

LTR Retrotransposons

LTR retrotransposons have direct LTRs that range from ~100 bp to over 5 kb in size. LTR retrotransposons are further sub-classified into the Ty1-copia-like (Pseudoviridae) , Ty3-gypsy-like (Metaviridae), and Pao-BEL-like groups based on both their degree of sequence similarity and the order of encoded gene products. Ty1-copia and Ty3-gypsy groups of retrotransposons are commonly found in high copy number (up to a few million copies per haploid nucleus) in animals, fungi, protista, and plants genomes. Pao-BEL like elements have so far only been found in animals. About 8% of the human genome and approximately 10% of the mouse genome are composed of the LTR transposons.

Ty1-copia Retrotransposons

Are abundant in species ranging from single-cell algae to bryophytes, gymnosperms, and angiosperms.

Ty3-gypsy Retrotransposons

Are also widely distributed, including both gymnosperms and angiosperms.

Non-LTR Retrotransposons

Consists of two sub-types, long interspersed nuclear elements (LINEs) and short interspersed nuclear elements (SINEs). They can also be found in high copy numbers (up to 250,000) in the plant species.

LINEs

Long interspersed repetitive elements or Long interspersed nuclear elements are a group of genetic elements that are found in large numbers in eukaryotic genomes. They are transcribed (or are the evolutionary remains of what was once transcribed) to an RNA using anRNA polymerase II promoter that resides inside the LINE. LINEs code for the enzyme reverse transcriptase, and many LINEs also code for an endonuclease (e.g. RNase H). The reverse transcriptase has a higher specificity for the LINE RNA than other RNA, and makes a DNA copy of the RNA that can be integrated into the genome at a new site.

The 5' UTR contains the promoter sequence, while the 3' UTR contains a polyadenylation signal (AATAAA) and a poly-A tail. Because

LINEs move by copying themselves (instead of moving, like transposons do), they enlarge the genome. The human genome, for example, contains about 500,000 LINEs, which is roughly 17% of the genome . Of these, approximately 7,000 are full-length, a small subset of which are capable of retrotransposition.

SINEs

Short interspersed repetitive elements or Short interspersed nuclear elements are short DNA sequences (<500 bases) that represent reverse-transcribed RNA molecules originally transcribed by RNA polymerase III into tRNA, rRNA, and other small nuclear RNAs. SINEs do not encode a functional reverse transcriptase protein and rely on other mobile elements for transposition. The most common SINEs in primates are called Alu sequences. Alu elements are 280 base pairs long, do not contain any coding sequences, and can be recognized by the restriction enzyme AluI (hence the name). With about 1,500,000 copies, SINEs make up about 11% of the human genome. While historically viewed as "junk DNA", recent research suggests that in some rare cases both LINEs and SINEs were incorporated into novel genes, so as to evolve new functionality.. The distribution of these elements has been implicated in some genetic diseases and cancers.

Composite SINE Transposons

Two SINES may act in concert to flank and mobilize an intervening single copy DNA sequence. This was reported for a 710 bp DNA sequence upstream of the bovine beta globin gene . The DNA arrangement forms a composite transposon whose presence has been confirmed by the complete bovine genomic sequence where the mobilized sequence may be found on bovine chromosome 15 in contigNW_001493315.1 nucleotides #1085432–1086142 and the originating sequence may be found on bovine chromosome 2 in contigNW_001501789.2 nucleotides #1096679–1097389. It is likely that similar composite transposons exist in other bovine genomic regions and other mammalian genomes. They could be detected with suitable algorithms.

Viroid

Viroids are plant pathogens that consist of a short stretch (a few hundred nucleobases) of highly complementary, circular, single-stranded RNA without the protein coat that is typical for viruses. The smallest discovered is a 220 nucleobase scRNA (small cytoplasmic RNA) associated with the rice yellow mottle sobemovirus (RYMV). In comparison, the genome of the smallest known viruses capable of causing an infection by themselves are around 2 kilobases in size. The human pathogen hepatitis D is similar to viroids.

Viroids were discovered and given this name by Theodor Otto Diener, a plant pathologist at the Agricultural Research Service in Maryland, in 1971. Viroid RNA does not code for any protein. The replication mechanism involves RNA polymerase II, an enzyme normally associated with synthesis of messenger RNA from DNA, which instead catalyses "rolling circle" synthesis of new RNA using the viroid's RNA as template. Some viroids are ribozymes, having catalytic properties which allow self-cleavage and ligation of unit-size genomes from larger replication intermediates.

The first viroid to be identified was the *potato spindle tuber viroid* (PSTVd). Some 33 species have been identified.

Primary Structure

1	CGGAACUAAA CUCGUGGUUC CUGUGGUUCA CACCUGACCU CCUGAGCAGA AAAGAAAAAA
61	GAAGGCGGCU CGGAGGAGCG CUUCAGGGAU CCCCGGGGAA ACCUGGAGCG AACUGGCAAA
121	AAAGGACGGU GGGGAGUGCC CAGCGGCCGA CAGGAGUAAU UCCCGCCGAA ACAGGGUUUU
181	CACCCUUCCU UUCUUCGGGU GUCCUUCCUC GCGCCCGCAG GACCACCCCU CGCCCCCUUU
241	GCGCUGUCGC UUCGGCUACU ACCCGGUGGA AACAACUGAA GCUCCCGAGA ACCGCUUUUU
301	CUCUAUCUUA CUUGCUUCGG GGCGAGGGUG UUUAGCCCUU GGAACCGCAG UUGGUUCCU

Secondary Structure

Taxonomy

- Family Pospiviroidae:
  - Genus *Pospiviroid*; type species: *Potato spindle tuber viroid*
  - Genus *Hostuviroid*; type species: *Hop stunt viroid*
  - Genus *Cocadviroid*; type species: *Coconut cadang-cadang viroid*

- o Genus *Apscaviroid*; type species: *Apple scar skin viroid*
- o Genus *Coleviroid*; type species: *Coleus blumei viroid 1.*

• Family Avsunviroidae;

- o Genus *Avsunviroid*; type species: *Avocado sunblotch viroid*
- o Genus *Pelamoviroid*; type species: *Peach latent mosaic viroid.*

Viroids and RNA Silencing

There has long been confusion over how viroids are able to induce symptoms in plants without encoding any protein products within their sequences. Evidence now suggests that RNA silencing is involved in the process. First, changes to the viroid genome can dramatically alter its virulence. This reflects the fact that any siRNAs produced would have less complementary base pairing with target messenger RNA. Secondly, siRNAs corresponding to sequences from viroid genomes have been isolated from infected plants. Finally, transgenicexpression of the noninfectious hpRNA of potato spindle tuber viroid develops all the corresponding viroid like symptoms.

This evidence indicates that when viroids replicate via a double stranded intermediate RNA, they are targeted by a dicer enzyme and cleaved into siRNAs that are then loaded onto the RNA-induced silencing complex. The viroid siRNAs actually contain sequences capable of complementary base pairing with the plant's own messenger RNAs and induction of degradation or inhibition of translation is what causes the classic viroid symptoms.

Satellite (Biology)

A satellite is a subviral agent composed of nucleic acid that depends on the co-infection of a host cell with a helper or master virus for its multiplication. When a satellite encodes the coat protein in which its nucleic acid is encapsidated it is referred to as a satellite virus. A satellite virus of mimivirus that inhibits the replication of its host has been termed a virophage.

Satellite viral particles should not be confused with satellite DNA.

Classification

Satellite viruses:

- Single-stranded RNA satellite viruses

- Subgroup 1: Chronic bee-paralysis virus associated satellite
- *Chronic bee-paralysis satellite virus*
- Subgroup 2: Tobacco necrosis virus satellite
- *Maize white line mosaic satellite virus*
- *Panicum mosaic satellite virus*
- *Tobacco mosaic satellite virus*
- *Tobacco necrosis satellite virus.*

Satellite nucleic acids;

- Single-stranded satellite DNAs
- *Tomato leaf curl virus satellite DNA*
- Double-stranded satellite RNAs
- *Saccharomyces cerevisiae M virus satellite*
- *Trichomonas vaginalis T1 virus satellite*
- Single-stranded satellite RNAs
- Subgroup 1: Large satellite RNAs
- *Arabis mosaic virus large satellite RNA*
- *Bamboo mosaic virus satellite RNA*
- *Chicory yellow mottle virus large satellite RNA*
- *Grapevine Bulgarian latent virus satellite RNA*
- *Grapevine fanleaf virus satellite RNA*
- *Myrobalan latent ringspot virus satellite RNA*
- *Tomato black ring virus satellite RNA*
- *Beet ringspot virus satellite RNA*
- Subgroup 2: Small linear satellite RNAs
- *Cucumber mosaic virus satellite RNA*
- *Cymbidium ringspot virus satellite RNA*
- *Pea enation mosaic virus satellite RNA*
- *Groundnut rosette virus satellite RNA*
- *Panicum mosaic virus small satellite RNA*
- *Peanut stunt virus satellite RNA*
- *Turnip crinkle virus satellite RNA*
- *Tomato bushy stunt virus satellite RNA, B10*
- *Tomato bushy stunt virus satellite RNA, B1*

- Subgroup 3: Circular satellite RNAs, or "virusoids"
- *Arabis mosaic virus small satellite RNA*
- *Cereal yellow dwarf virus-RPV satellite RNA*
- *Chicory yellow mottle virus satellite RNA*
- *Lucerne transient streak virus satellite RNA*
- *Solanum nodiflorum mottle virus satellite RNA*
- *Subterranean clover mottle virus satellite RNA*
- *Tobacco ringspot virus satellite RNA*
- *Velvet tobacco mottle virus satellite RNA.*

Vault RNA

ault RNAs (vRNA) are an RNA family found as part of the *vault ribonucleoprotein complex* first discovered in 1986. The complex consists of a *major vault protein* (MVP), two *minor vault proteins* (VPARP and TEP1), and several small untranslated RNA molecules. Each vault particle contains 8-16 vRNA molecules. The vault complex has been linked to drug resistance. Cryo-electron microscopy has revealed the vRNA resides internally close to the ends of the vault caps. The position of vRNA suggests that it could interact with both the interior and exterior of the vault particle. It has been shown that TEP1 is involved in stabilising the vRNA.

Species Distribution

Vault RNAs have been found in:

- Human
- Rodents
- Bullfrog.

Vaults but not the vRNA have also been isolated from:

- sea urchin
- Dictyostelium discoideum
- Acanthamoeba.

Role in Multidrug Resistance

Vault RNAs, in conjunction with the vault ribonucleoprotein complex, have long been connected with drug resistance, though the mechanisms involved have only recently been determined. Vault non-coding RNAs produce small vault RNAs (svRNAs) through a DICER mechanism which then operate in similar manner to miRNAs: an svRNA binds an argonaute protein and downregulates expression of CYP3A4, an enzyme involved in drug metabolism.

In Translation

Messenger RNA (mRNA) carries information about a protein sequence to the ribosomes, the protein synthesis factories in the cell. It is coded so that every three nucleotides (a codon) correspond to one amino acid. In eukaryotic cells, once precursor mRNA (pre-mRNA) has been transcribed from DNA, it is processed to mature mRNA. This removes its introns—non-coding sections of the pre-mRNA. The mRNA is then exported from the nucleus to the cytoplasm, where it is bound to ribosomes and translated into its corresponding protein form with the help of tRNA. In prokaryotic cells, which do not have nucleus and cytoplasm compartments, mRNA can bind to ribosomes while it is being transcribed from DNA. After a certain amount of time the message degrades into its component nucleotides with the assistance of ribonucleases. Transfer RNA (tRNA) is a small RNA chain of about 80 nucleotides that transfers a specific amino acid to a growing polypeptide chain at the ribosomal site of protein synthesis during translation. It has sites for amino acid attachment and an anticodon region for codon recognition that binds to a specific sequence on the messenger RNA chain through hydrogen bonding.

Ribosomal RNA (rRNA) is the catalytic component of the ribosomes. Eukaryotic ribosomes contain four different rRNA molecules: 18S, 5.8S, 28S and 5S rRNA. Three of the rRNA molecules are synthesized in the nucleolus, and one is synthesized elsewhere. In the cytoplasm, ribosomal RNA and protein combine to form a nucleoprotein called a ribosome. The ribosome binds mRNA and carries out protein synthesis. Several ribosomes may be attached to a single mRNA at any time. rRNA is extremely abundant and makes up 80% of the 10 mg/ml RNA found in a typical eukaryotic cytoplasm.

Transfer-messenger RNA (tmRNA) is found in many bacteria and plastids. It tags proteins encoded by mRNAs that lack stop codons for degradation and prevents the ribosome from stalling.

Regulatory RNAs

Several types of RNA can downregulate gene expression by being complementary to a part of an mRNA or a gene's DNA. Micro RNAs (mi RNA; 21-22 nt) are found in eukaryotes and act through RNA interference (RNAi), where an effector complex of miRNA and enzymes can break down mRNA which the miRNA is complementary to, block the mRNA from being translated, or accelerate its degradation. Whilesmall interfering RNAs (siRNA; 20-25 nt) are often produced by breakdown of viral RNA, there are also endogenous sources of siRNAs.

siRNAs act through RNA interference in a fashion similar to miRNAs. Some miRNAs and siRNAs can cause genes they target to bemethylated, thereby decreasing or increasing transcription of those genes. Animals have Piwi-interacting RNAs (piRNA; 29-30 nt) which are active in germline cells and are thought to be a defence against transposons and play a role in gametogenesis.

Many prokaryotes have CRISPR RNAs, a regulatory system similar to RNA interference. Antisense RNAs are widespread; most downregulate a gene, but a few are activators of transcription. One way antisense RNA can act is by binding to an mRNA, forming double-stranded RNA that is enzymatically degraded. There are many long noncoding RNAs that regulate genes in eukaryotes, one such RNA is Xist which coats one X chromosome in female mammals and inactivates it. An mRNA may contain regulatory elements itself, such as riboswitches, in the 5' untranslated region or 3' untranslated region; these cis-regulatory elements regulate the activity of that mRNA. The untranslated regions can also contain elements that regulate other genes.

In RNA Processing

Many RNAs are involved in modifying other RNAs. Introns are spliced out of pre-mRNA byspliceosomes, which contain several small nuclear RNAs (snRNA), or the introns can be ribozymes that are spliced by themselves. RNA can also be altered by having its nucleotides modified to other nucleotides than A, C, G and U. In eukaryotes, modifications of RNA nucleotides are generally directed by small nucleolar RNAs (snoRNA; 60-300 nt), found in the nucleolusand cajal bodies. snoRNAs associate with enzymes and guide them to a spot on an RNA by basepairing to that RNA. These enzymes then perform the nucleotide modification. rRNAs and tRNAs are extensively modified, but snRNAs and mRNAs can also be the target of base modification.

RNA Genomes

Like DNA, RNA can carry genetic information. RNA viruses have genomes composed of RNA, and a variety of proteins encoded by that genome. The viral genome is replicated by some of those proteins, while other proteins protect the genome as the virus particle moves to a new host cell.Viroids are another group of pathogens, but they consist only of RNA, do not encode any protein and are replicated by a host plant cell's polymerase.

In Reverse Transcription

Reverse transcribing viruses replicate their genomes by reverse transcribing DNA copies from their RNA; these DNA copies are then transcribed to new RNA. Retrotransposons also spread by copying DNA and RNA from one another, and telomerase contains an RNA that is used as template for building the ends of eukaryotic chromosomes.

Double-stranded RNA

Double-stranded RNA (dsRNA) is RNA with two complementary strands, similar to the DNA found in all cells. dsRNA forms the genetic material of some viruses (double-stranded RNA viruses). Double-stranded RNA such as viral RNA or siRNA can trigger RNA interference ineukaryotes, as well as interferon response in vertebrates.

Key Discoveries in RNA Biology

Research on RNA has led to many important biological discoveries and numerous Nobel Prizes. Nucleic acids were discovered in 1868 by Friedrich Miescher, who called the material 'nuclein' since it was found in the nucleus. It was later discovered that prokaryotic cells, which do not have a nucleus, also contain nucleic acids. The role of RNA in protein synthesis was suspected already in 1939. Severo Ochoawon the 1959 Nobel Prize in Medicine (shared with Arthur Kornberg) after he discovered an enzyme that can synthesize RNA in the laboratory. Ironically, the enzyme discovered by Ochoa (polynucleotide phosphorylase) was later shown to be responsible for RNA degradation, not RNA synthesis.

The sequence of the 77 nucleotides of a yeast tRNA was found by Robert W. Holley in 1965, winning Holley the 1968 Nobel Prize in Medicine (shared with Har Gobind Khorana and Marshall Nirenberg). In 1967, Carl Woese hypothesized that RNA might be catalytic and suggested that the earliest forms of life (self-replicating molecules) could have relied on RNA both to carry genetic information and to catalyse biochemical reactions—an RNA world.

During the early 1970's, retroviruses and reverse transcriptase were discovered, showing for the first time that enzymes could copy RNA into DNA (the opposite of the usual route for transmission of genetic information). For this work, David Baltimore, Renato Dulbecco and Howard Temin were awarded a Nobel Prize in 1975. In 1976, Walter Fiers and his team determined the first complete nucleotide sequence of an RNA virus genome, that of bacteriophage MS2.

In 1977, introns and RNA splicing were discovered in both mammalian viruses and in cellular genes, resulting in a 1993 Nobel to Philip Sharpand Richard Roberts. Catalytic RNA molecules (ribozymes) were discovered in the early 1980s, leading to a 1989 Nobel award to Thomas Cech and Sidney Altman. In 1990 it was found in petunia that introduced genes can silence similar genes of the plant's own, now known to be a result of RNA interference. At about the same time, 22 nt long RNAs, now called microRNAs, were found to have a role in the development of *C. elegans*. Studies on RNA interference gleaned a Nobel Prize for Andrew Fire and Craig Mello in 2006, and another Nobel was awarded for studies on transcription of RNA to Roger Kornberg in the same year. The discovery of gene regulatory RNAs has led to attempts to develop drugs made of RNA, such as siRNA, to silence genes.

Protein Biosynthesis

Protein synthesis is the process in which cells build proteins. The term is sometimes used to refer only to protein translation but more often it refers to a multi-step process, beginning with amino acid synthesis and transcription of nuclear DNAinto messenger RNA, which is then used as input to translation.

The cistron DNA is transcribed into a variety of RNA intermediates. The last version is used as a template in synthesis of a polypeptide chain. Proteins can often be synthesized directly from genes by translating mRNA. When a protein needs to be available on short notice or in large quantities, a protein precursor is produced. Aproprotein is an inactive protein containing one or more inhibitory peptides that can be activated when the inhibitory sequence is removed by proteolysis duringposttranslational modification. A preprotein is a form that contains a signal sequence (an N-terminal signal peptide) that specifies its insertion into or through membranes; i.e., targets them for secretion. The signal peptide is cleaved off in theendoplasmic reticulum. Preproproteins have both sequences (inhibitory and signal) still present.

For synthesis of protein, a succession of tRNA molecules charged with appropriate amino acids have to be brought together with an mRNA molecule and matched up by base-pairing through their anticodons with each of its successive codons. The amino acids then have to be linked together to extend the growing protein chain, and the tRNAs, relieved of their burdens, have to be released. This whole complex of processes is carried out by a giant multimolecular machine, the ribosome, formed of two main chains of RNA, called ribosomal RNA (rRNA), and more than 50 different proteins. This molecular juggernaut latches onto the end of an mRNA molecule and then trundles along it, capturing loaded tRNA molecules and stitching together the amino acids they carry to form a new protein chain.

Protein biosynthesis, although very similar, is different for prokaryotes and eukaryotes.

Bibliography

Alberts B, Johnson A: *Molecular Biology of the Cell,* New York: Garland Science, 2002.

Altered Fates: *Gene Therapy and the Retooling of Human Life,* W. W. Norton and Company, New York, 1995.

Ayala, F. J.: *The Genetic Structure of Populations,* W.H. Freeman & Co., San Francisco, California, 1977.

Bajema, C. J.: *Natural Selection in Human Populations, the Measurement of Ongoing Genetic Evolution in Contemporary Societies*: New York, Wiley, 1971.

Balkwill, Dr. Fran: *DNA Is Here to Stay,* Carolrhoda Books, Inc., Minneapolis, 1993.

Balkwill, Fran: *DNA Is Here to Stay,* Carolrhoda Books, Inc., Minneapolis, 1993.

Banerjee, S : *Cell Biochemistry,* Dominant Pub, Delhi, 2009.

Bast, R.C. Jr.: *Cancer Medicine,* Decker, Hamilton, 2000.

Beckner, M.: *The Biological Way of Thought,* University of California Press, Berkeley and Los Angeles, 1968.

Berg, Paul, and Singer, Maxine: *Dealing with Genes: The Language of Heredity. Mill Valley,* University Science Books, CA, 1992.

Bernard, C.: *An Introduction to the Study of Experimental Medicine,* Dover Publications, New York, 1957.

Birch, L. C.: *The Distribution and Abundance of Animals,* University of Chicago Press, Chicago, Illinois, 1954

Bradley, W.G.: *Neurology in Clinical Practice,* Butterworth Heinemann, Boston, 2000.

Brookes, Martin: *Get a Grip on Genetics, Time Life Books,* East Sussex, England, 1998.

Cavalli-Sforza, L.L., P. Menozzi: *The History and Geography of Human Genes.* Princeton: Princeton University Press, 1994.

Credi, A.: *Molecular Devices and Machines: a Journey into the Nano World,* Wiley-VCH, Weinheim, 2003.

Curtis, M.: *The Geometry of DNA: A Structural Revision.* London: Blue Gallery, 1996.

Dabholkar, A.R. : *General Plant Breeding*, Concept, Delhi, 2006.

Dando, M.: *Biological Warfare in the 21st Century: Biotechnology and the Proliferation of Biological Weapons*. London: Brassey's, 1994.

Darlington, C. D., and Mather, K.: *The Elements of Genetics*, London, Allen and Unwin, 1949.

David M.: *Basic Principles and Calculations in Chemical Engineering*, Prentice-Hall, 1996.

Dawkins, R.: *The Selfish Gene*. Oxford: Oxford University Press, 1976.

DeVita, V.T.: *Cancer: Principles and Practice of Oncology*, Lippincott, Philadelphia, 2005.

Diaz, E.: *Microbial Biodegradation: Genomics and Molecular Biology*, Caister Academic Press, UK, 2008.

Dobzhanshy, T.: *Genetics of the Evolutionary Process*, Columbia University Press, New York, 1970.

Easty, D. M.: *Cell Biology: Reading*, Addison- Wesley, Mass, 1970.

Fisher, R. A.: *The Genetical Theory of Natural Selection*, Clarendon Press, Oxford, 1930.

Fritz, A.: *International Classification of Diseases for Oncology*, World Health Organization, Geneva, 2000.

Fukuyama F.: *Our Posthuman Future: Consequences of the Biotechnology Revolution*, Farrar, Straus and Giroux, NewYork, 2002.

Gall, Joseph G.: *Landmark Papers in Cell Biology*, Cold Spring Harbor Laboratory Press, Plainview, NY, 2001.

George S. Paul: *Beyond Humanity: Cyber Evolution and Future Minds*, Charles River Media, Roackland, 1996.

Gillis, Justin: *Drug Firms, Gene Labs to Map Genetic Code*, The Daily News, Longview, WA, 1999.

Glover, D. M., and Hames, B. D.: *Genes and Embryos*: New York, Oxford University Press, 1989.

Glut, D.F.: *The Frankenstein Legend: A Tribute to Mary Shelley and Boris Karloff*. Metachen, New Jersey: The Scarecrow Press, 1973.

Griffiths A.J.F.: *Introduction to Genetic Analysis*, W.H. Freeman and Company, New York, USA, 2005.

Groom, M.J., Meffe, G.K. : *Principles of Conservation Biology*, Sunderland, MA: Sinauer Associates, 2006.

Guyton, A.C.; Hall, J.E.: *Textbook of Medical Physiology*, Saunders, Philadelphia, 2006.

Halacy, D.S., Jr.: *Genetic Revolution, Shaping Life for Tomorrow*, Harper & Row, Publishers, New York, 1974.

Hamerton, J. L.: *Human Cytogenetics*, Academic Press, New York, 1971.

Hanley, R.: *Is Data Human? The Metaphysics of Star Trek*, Boxtree, London, 1998.

Hartl D, Jones E: *Genetics: Analysis of Genes and Genomes,* Jones & Bartlett, 2005.

Hartl, Daniel L.: *Basic Genetics*, Jones and Bartlett Publishers, Boston, 1991.

Haubrich, W.S.: *Bockus Gastroenterology*, Saunders, Philadelphia, 1995.

Heider, J. and Rabus, R: *Microbial Biodegradation: Genomics and Molecular Biology*, Caister Academic Press, UK, 2008.

Herrero A and Flores E,: *The Cyanobacteria: Molecular Biology, Genomics and Evolution* (1st ed.). Caister Academic Press, 2008.

Holtz, Robert D.; William, Kovacs D.: *An Introduction to Geotechnical Engineering*, Prentice Hall, UK, 1981.

Jangir, M.L. : *Cell Biology : Fundamentals and Applications*, Agrobios, Delhi, 2009.

Jonoska, N.: *Self-Assembling DNA Graphs, DNA-Based Computers VIII*, Springer-Verlag, Berlin, 2003.

Kelves, D.: *In the Name of Eugenics: Genetics and the Uses of Human Heredity*. Harmondsworth: Penguin.

King, R.C.; Stansfield, W.D.: *A Dictionary of Genetics*, Oxford Univ. Press, New York, 2002.

Klotz, J. W.: *Genes, Genesis, and Evolution*: St. Louis, Mo., Concordia Publishing Co., 1970.

Landa, Norbert: *Ingenious Genes Microexplorers*, Barron's, Hauppauge, New York, 1997.

Lewis, Ricki: *Human Genetics: Concepts and Applications,* Boston, Mass: McGraw Hill. 1994.

Lewontin, R.C.: *The Doctrine of DNA: Biology as Ideology*. Harmondsworth: Penguin, 1993.

Lindee, M.S. : *The DNA Mystique: The Gene as a Cultural Icon*. New York: W.H. Freeman, 1995.

Luca Cavalli-Sforza, Luigi: *The History and Geography of Human Genes*. Princeton University Press, 1996.

Lumsden C., and E. Wilson: *Genes, Mind and Culture: The Coevolutionary Process*. Cambridge, MA: Harvard University Press, 1981.

Lyon, Jeff and Gorner, Peter. Altered Fates: *Gene Therapy and the Retooling of Human Life*, W. W. Norton and Company, New York, 1995.

Margulis, L.: *Symbiosis in Cell Evolution*. San Francisco: W.H. Freeman, 1981.

Mayr, E.: *Change of Genetic Environment and Evolution*, Allen and Unwin, London, 1954.

Michael R.: Cummings. *Essentials of Genetics*, Prentice Hall, New Jersey, 1996.

Migloni, G.S.: *Dictionary of Plant Genetics and Molecular Biology*, Hawthorne Press, New York, 1998.

Mishra, Shubhrata R. : *Textbook of Mycology*, Discovery Publishing House, Delhi, 2010.

Nottingham, S.F.: *Eat Your Genes: How Genetically Modified Food is Entering Our Diet*, Zed Books, London, 1998.

Prakash, Chandra Prakash : *Cell Physiology*, Discovery, Delhi, 2010.

Rietman, Ed.: *Molecular Engineering of Nanosystems*, Springer, New York, 2001.

Robin J. Heyden: *Biology: Exploring Life*, Pearson Prentice Hall, Boston, Massachusetts, 2006.

Robin Whyman: *Applied Organometallic Chemistry and Catalysis*, Oxford University Press, Delhi, 2005.

Roobeek, A: *The Biotechnology Revolution?*, Blackwell, Oxford, 1995.

Rorvik, D.M.: *In His Image: The Cloning of a Man*. New York: J.B. Lippincott, 1978.

Scriver, C.R.: *The Metabolic and Molecular Basis of Inherited Disease*, McGraw Hill, New York, 2001.

Stebbins, G. L.: *Darwin to DNA, Molecules to Humanity*, W. H. Freeman, San Francisco, 1982.

Steen, R.G.: *DNA and Destiny: Nature and Nurture in Human Behaviour*. New York and London: Plenum Press, 1996.

Sturtevant, Alfred: *History of Genetics*, Harper and Row, New York, 1965.

Stwertka, Eve and Albert: *Genetic Engineering*, Franklin Watts, New York, 1982.

Suzuki, D., and Knudtson, P.: *Genethics: The Clash Between the New Genetics and Human Values*, Harvard University Press, Cambridge, Mass, 1989.

Vasudev Bhatnagar: *Cell Science and Technology*, Campus Books, Delhi, 2009.

Watson, J. D.: *The Double Helix*, Antheneum, New York, 1968.

Wilson, E. O., and Lumden, C.: *Genes, Mind, and Culture: The Evolutionary Process*, Harvard University Press, Cambridge, Mass, 1981.

Wilson, E. O., and Lumden, C.: *Genes, Mind, and Culture: The Evolutionary Process*: Cambridge, Mass., Harvard University Press, 1981.

Wilson, I.: *The Blood and The Shroud*. London: Weidenfeld & Nicolson, 1998.

Winston, R.: *The Future of Genetic Manipulation*, Phoenix, London, 1997.

Wright, L.: *Twins: Genes, Environment and the Mystery of Human Identity*. London: Weidenfeld & Nicolson, 1997.

Index

G

H

OOO